DAVID CHRISTIAN

Origin Story

A Big History of Everything

PENGUIN BOOKS

PENGUIN BOOKS

UK | USA | Canada | Ireland | Australia
India | New Zealand | South Africa

Penguin Books is part of the Penguin Random House group of companies
whose addresses can be found at global.penguinrandomhouse.com.

First published in the United States of America
by Little, Brown and Company 2018
First published in Great Britain by Allen Lane 2018
Published in Penguin Books 2019
001

Printed and bound in Great Britain by Clays Ltd, Elcograf S.p.A.

A CIP catalogue record for this book is available from the British Library

ISBN: 978–0–141–98302–8

Origin Story

'Long-haul science with wit and oomph' *Nature*

'Sets human history in the light of the wider trends that have shaped the universe . . . This does indeed cover everything from supernovae and slime moulds to slavery and steam engines, taking readers through a series of eight thresholds eventually leading to the Anthropocene and the challenge of climate change'
Matthew Reisz, *Times Higher Education*

'In *Origin Story*, David Christian has found a spectacular way to use history to put order in the entire set of our knowledge about the world. This is a wonderful achievement'
Carlo Rovelli, author of *Seven Brief Lessons on Physics* and *The Order of Time*

'A remarkable book that puts us self-important humans in our proper place in the cosmos, yet also explains why the story of human culture and knowledge – what Christian calls collective learning – matters for understanding our present world and shaping its future'
Merry Wiesner-Hanks, President of the World History Association

'*Origin Story* is a majestic distillation of our current understanding of the birth and development of the universe, of the solar system, of the oceans, of mountains and minerals, of all life on earth and of the driving dynamics of human culture and achievement. All of this in just over 300 pages of captivating prose that weaves together innumerable insights from dozens of disciplines in the sciences, arts and humanities. With fascinating ideas on every page and the page-turning energy of a good thriller, this is a landmark work that comes at a time when it has never been more important for humanity as a whole to have a clearer, more informed understanding of our place on earth and of the earth's place in the cosmos' Sir Ken Robinson, author of *The Element*

'A wonderful story . . . Mr Christian tells this story very well, providing, in effect, a short course in modern science. This is a brief history of the universe, and an excellent one' David Wootton, *Wall Street Journal*

'The most powerful example of interdisciplinary scholarship that I know of' Fareed Zakaria

ABOUT THE AUTHOR

David Christian is a distinguished professor in history at Macquarie University in Australia and the co-founder, with Bill Gates, of The Big History Project, which has built a free online syllabus on the history of the universe and is taught in schools all over the world. He is also co-creator of Macquarie University Big History School, which provides online courses in big history for primary and high school students. He received his PhD from the University of Oxford. He has delivered keynotes at conferences around the world including at the Davos World Economic Forum, and his TED Talk on the history of the Universe has been viewed over 7 million times.

Contents

Contents

PART IV: THE FUTURE

Preface

We tell stories to make sense of things. It's in our blood.
— Lia Hills, "Return to the Heart"

The idea of a modern origin story is in the air. For me, it began with a course on the history of everything that I first taught at Macquarie University in Sydney in 1989. I saw that course as a way of getting at the history of humanity. At the time, I taught and researched Russian and Soviet history. But I worried that teaching a national or imperial history (Russia was both nation and empire) conveyed the subliminal message that humans are divided, at the most fundamental level, into competing tribes. Was that a helpful message to teach in a world with nuclear weapons? As a schoolboy during the Cuban missile crisis, I vividly remember thinking we were on the verge of an apocalypse. Everything was about to be destroyed. And I remember wondering if there were kids "over there" in the Soviet Union who were equally scared. After all, they, too, were humans. As a child, I had lived in Nigeria. That gave me a strong sense of a single, extraordinarily diverse human community, a feeling that was confirmed when, as a teenager, I went to Atlantic College, an international school in South Wales.

Several decades later, as a professional historian, I began to think about how to teach a unified history of humanity. Could I teach about the heritage shared by all humans and tell that story with some of the grandeur and awe of the great national histories? I became convinced that we needed a story in which our Paleolithic ancestors and Neolithic farmers could play as important a role as the rulers, conquerors, and emperors who have dominated so much historical scholarship.

Eventually, I understood that these were not original ideas. In 1986, the great world historian William McNeill argued that writing histories of "the triumphs and tribulations of humanity as a whole" was "the moral duty of the historical profession in our time."[1] Even earlier, but in the same spirit, H. G. Wells wrote a history of humanity as a response to the carnage of World War I.

> There can be no peace now, we realize, but a common peace in all the world; no prosperity but a general prosperity. But there can be no common peace and prosperity without common historical ideas....With nothing but narrow, selfish, and conflicting nationalist traditions, races and peoples are bound to drift towards conflict and destruction.[2]

Wells understood something else, too: If you want to teach the history of humanity, you probably need to teach the history of everything. That's why his *Outline of History* turned into a history of the universe. To understand the history of humanity, you have to understand how such a strange species evolved, which means learning about the evolution of life on planet Earth, which means learning about the evolution of planet Earth, which means learning about the evolution of stars and planets, which means knowing about the evolution of the universe. Today, we can tell that story with a precision and scientific rigor that was unthinkable when Wells wrote.

Wells was looking for unifying knowledge—knowledge that links disciplines as well as peoples. All origin stories unify knowledge, even the origin stories of nationalist historiography. And the most capacious of them can lead you across many time scales and through many concentric circles of understanding and identity, from the self to the family and clan, to a nation, language group, or religious affiliation, to the huge circles of humanity and life, and eventually to the idea that you are part of an entire universe or cosmos.

But in recent centuries, increasing cross-cultural contacts have shown how embedded all origin stories and religions are in local customs and environments. That is why globalization and the spread of new ideas corroded faith in traditional knowledge. Even true believers began to see that there were multiple origin stories that said very different things. Some people responded with aggressive, even violent, defenses of their own religious, tribal, or national traditions. But many simply lost faith and conviction, and along with them, they lost their bearings, their sense of their place in the universe. That loss of faith helps explain the pervasive *anomie,* the feeling of aimlessness, meaninglessness, and sometimes even despair that shaped so much literature, art, philosophy, and scholarship in the twentieth century. For many, nationalism offered some sense of belonging, but in today's globally connected world, it is apparent that nationalism divides humanity even as it connects citizens within a particular country.

I have written this book in the optimistic belief that we moderns are not doomed to a chronic state of fragmentation and meaninglessness. Within the creative hurricane of modernity, there is emerging a new, global origin story that is as full of meaning, awe, and mystery as any traditional origin story but is based on modern scientific scholarship across many disciplines.[3] That story is far from complete, and it may need to incorporate the insights of older origin stories about how to live well and how to live sustainably. But it is worth knowing, because it draws on a global heritage of carefully tested information and knowledge and it is the first origin story to embrace human societies and cultures from around the world. It is a collective global project, a story that should work as well in Buenos Aires as in Beijing, as well in Lagos as in London. Today, many scholars are engaged in the exciting task of building and telling this modern origin story, looking for the guidance and sense of shared purpose that it may provide, like all origin stories, but for today's globalized world.

My own attempts to teach a history of the universe began in 1989. In 1991, as a way to describe what I was doing, I started using the term *big history*.[4] Only as the story slowly came into focus did I realize that I was trying to tease out the main lines of an emerging global origin story. Today, big history is being taught in universities in many different parts of the world, and through the Big History Project, it is also being taught in thousands of high schools.

We will need this new understanding of the past as we grapple with the profound global challenges and opportunities of the twenty-first century. This book is my attempt to tell an up-to-date version of this huge, elaborate, beautiful, and inspiring story.

ORIGIN
STORY

Introduction

The forms that come and go — and of which your body is but one — are the flashes of my dancing limbs. Know Me in all, and of what shall you be afraid?

— IMAGINED WORDS OF THE HINDU GOD
SHIVA, FROM JOSEPH CAMPBELL, *THE HERO
WITH A THOUSAND FACES*

Utterly impossible as are all these events they are probably as like those which may have taken place as any others which never took person at all are ever likely to be.

— JAMES JOYCE, *FINNEGANS WAKE*

We arrive in this universe through no choice of our own, at a time and place not of our choosing. For a few moments, like cosmic fireflies, we will travel with other humans, with our parents, with our sisters and brothers, with our children, with friends and enemies. We will travel, too, with other life-forms, from bacteria to baboons, with rocks and oceans and auroras, with moons and meteors, planets and stars, with quarks and photons and supernovas and black holes, with slugs and cell phones, and with lots and lots of empty space. The cavalcade is rich, colorful, cacophonous, and mysterious, and though we humans will eventually leave it, the cavalcade will move on. In the remote future, other travelers will join and leave the cavalcade. Eventually, though, the cavalcade will thin out. Gazillions of years from today, it will fade away like a ghost at dawn, dissolving into the ocean of energy from which it first appeared.

What is this strange crowd we travel with? What is our place in the cavalcade? Where did it set out from, where is it heading, and how will it finally fade away?

Today, we humans can tell the story of the cavalcade better than ever before. We can determine with remarkable accuracy what lurks out there, billions of light-years from Earth, as well as what was going on billions of years ago. We can do this because we have so many more pieces of the jigsaw puzzle of knowledge, which makes it easier to figure out what the whole picture may look like. This is an astonishing, and very recent, achievement. Many of the pieces of our origin story fell into place during my own lifetime.

We can build these vast maps of our universe and its past partly because we have large brains, and, like all brainy organisms, we use our brains to create internal maps of the world. These maps provide a sort of virtual reality that helps us find our way. We can never see the world directly in all its detail; that would require a brain as big as the universe. But we can create simple maps of a fantastically complicated reality, and we know that those maps correspond to important aspects of the real world. The conventional diagram of the London Underground ignores most of the twists and turns, but it still helps millions of travelers get around the city. This book offers a sort of London Underground map of the universe.

What makes humans different from all other brainy species is language, a communication tool that is extraordinarily powerful because it allows us to share our individual world maps and, in so doing, form maps much larger and more detailed than those created by an individual brain. Sharing also allows us to test the details of our maps against millions of other maps. In this way, each group of humans builds up an understanding of the world that combines the insights, ideas, and thoughts of many people over thousands of years and many generations. Pixel by pixel, through this process of collective learning, humans have built increasingly rich maps of the universe during the two hundred thousand years of our existence as a species. What this means is that one small part of the universe is beginning to look at itself. It's as if the universe were slowly opening an eye after a long sleep. Today, that eye is seeing more and more, with the help of global exchanges of ideas and information; the precision and rigor of modern science; new research instruments, from atom-smashing particle colliders to space-based telescopes; and networks of computers with colossal number-crunching powers.

The story these maps tell us is the grandest story you can imagine.

As a child, I could not make sense of anything unless I could place it on some sort of map. Like many people, I struggled to link the isolated fields I studied. Literature had nothing to do with physics; I could see no connection between philosophy and biology, or religion and mathematics, or economics and ethics. I kept looking for a framework, a sort of world map of the different continents and islands of human knowledge; I wanted to be able to see how they all fitted together. Traditional religious stories never quite worked for me because, having lived in Nigeria as a child, I'd learned very early that different religions offer different, and often contradictory, frameworks for understanding how the world came to be as it is.

Today, a new framework for understanding is emerging in our globalized world. It is being built, developed, and propagated collectively by thousands of people from multiple scholarly fields and in numerous countries. Linking these insights can help us see things that we cannot see from within the boundaries of a particular discipline; it lets us view the world from a mountaintop instead of from the ground. We can see the links connecting the various scholarly landscapes, so we can think more deeply about broad themes such as the nature of complexity, the nature of life, even the nature of our own species! After all, at present we study humans through many different disciplinary lenses (anthropology, biology, physiology, primatology, psychology, linguistics, history, sociology), but specialization makes it difficult for any individual to stand back far enough to see humanity as a whole.

The search for origin stories that can link different types of knowledge is as old as humanity. I like to imagine a group of people sitting around a fire as the sun was setting forty thousand years ago. I picture them on the southern shore of Lake Mungo, in the Willandra Lakes Region of New South Wales, where the oldest human remains in Australia have been found. Today, it is the home of the Paakantji, Ngyiampaa, and Mutthi Mutthi people, but we know that their ancestors lived in this region for at least forty-five thousand years.

In 1992, the remains of an ancestor (referred to as Mungo 1) discovered by archaeologists in 1968 were finally returned to the local Aboriginal community. This person was a young woman who had been partially cremated.[1] Half a kilometer away, remains were found of another person (Mungo 3), probably a man, who died at about age fifty. He had suffered from arthritis and severe dental erosion, probably caused by drawing fibers through his teeth to make nets or cords. His body had been buried with care and reverence and sprinkled with powdered red ocher brought from two hundred kilometers away. Mungo Man was returned to Lake Mungo in November 2017.

Both people died about forty thousand years ago, when the Willandra lakes, which are now dry, were full of water, fish, and shellfish and attracted multitudes of birds and animals that could be hunted or trapped.[2] Life was pretty good around Lake Mungo when they were alive.

In my imagined twilight conversations around the fire, there are girls and boys, older men and women, and parents and grandparents, some wrapped in animal furs and cradling babies. Children are chasing one another at the edge of the lake while adults are finishing a meal of mussels, freshly caught fish and yabbies, and wallaby steak. Slowly, the conversation becomes serious and is taken over by one of the older people. As on many long summer days and cold winter nights, the older people are retelling what they have learned from their ancestors and teachers. They are asking the sort of questions that have always fascinated me: How did the landscape, with its hills and lakes, its valleys and ravines, take shape? Where do the stars come from? When did the first humans live, and where did they come from? Or have we always been here? Are we related to goannas and wallabies and emus? (The answer of both the Lake Mungo people and modern science to that last question is a decisive "Yes!") The storytellers are teaching history. They are telling stories about how our world was created by powerful forces and beings in the distant past.

Told over many nights and days, their stories describe the big paradigm ideas of the Lake Mungo people. These are the ideas with long legs, the ideas that can stay the course. They fit together to form a vast mosaic of information about the world. Some of the children may find parts of the stories too complex and subtle to take in at first hearing. But they hear the stories many times in different tellings, and they get used to them and to the deep ideas inside the stories. As the children get older, the stories get under their skin. They come to know them intimately and better appreciate their beauty and their subtler details and meanings.

As they talk about the stars, the landscape, the wombats and the wallabies, and the world of their ancestors, the teachers build a shared map of understanding that shows members of the community their place in a rich, beautiful, and sometimes terrifying universe: *This* is what you are; *this* is where you came from; *this* is who existed before you were born; *this* is the whole thing of which you are a small part; *these* are the responsibilities and challenges of living in a community of others like yourself. The stories have great power because they are trusted. They *feel* true because they are based on the best knowledge passed down by ancestors over many generations. They have been checked and rechecked for accuracy, plausibility, and coherence using the rich knowledge of people, of stars, of landscapes, of plants and animals available to the Mungo community and to their ancestors and neighbors.

We can all benefit from the maps our ancestors created. The great French sociologist Émile Durkheim insisted that the maps lurking within origin stories and religions were fundamental to our sense of self. Without them, he argued, people could fall into a sense of despair and meaninglessness so profound, it might drive them to suicide. No wonder almost all societies we know of have put origin stories at the heart of education. In Paleolithic societies, students learned origin stories from their elders, just as later scholars learned the core stories of Christianity, Islam, and Buddhism in the universities of Paris, Oxford, Baghdad, and Nalanda.

Yet, curiously, modern secular education lacks a confident origin story that links all domains of understanding. And that may help explain why the sense of disorientation, division, and directionlessness that Durkheim described is palpable everywhere in today's world, in Delhi or Lima as much as in Lagos or London. The problem is that in a globally connected world, there are so many local origin stories competing for people's trust and attention that they get in one another's way. So most modern educators focus on parts of the story, and students learn about their world discipline by discipline. People today learn about things our Lake Mungo ancestors had never heard of, from calculus to modern history to how to write computer code. But, unlike the Lake Mungo people, we are rarely encouraged to assemble that knowledge into a single, coherent story in the way that globes in old-fashioned classrooms linked thousands of local maps into a single map of the world. And that leaves us with a fragmented understanding of both reality and the human community to which we all belong.

A Modern Origin Story

And yet ... in bits and pieces, a modern origin story is emerging. Like the stories told at Lake Mungo, our modern origin story has been assembled by ancestors and tested and checked over many generations and millennia.

It is different, of course, from most traditional origin stories. This is partly because it has been built not by a particular region or culture but by a global community of more than seven billion people, so it pools knowledge from all parts of the world. This is an origin story for all modern humans, and it builds on the global traditions of modern science.

Unlike many traditional origin stories, the modern origin story lacks a creator god, though it has energies and particles as exotic as the pantheons of many traditional origin stories. Like the origin stories of Confucianism or early Buddhism, the modern story is about a universe that just is. Any sense of meaning comes not from the universe, but from us humans. "What's the meaning of the universe?" asked Joseph Campbell, a scholar of myth and religion. "What's the meaning of a flea? It's just there, that's it, and your own meaning is that you're there."[3]

The world of the modern origin story is less stable, more turbulent, and much larger than the worlds of many traditional origin stories. And those qualities point to the limitations of the modern origin story. Though global in its reach, it is very recent and it has the rawness and some of the blind spots of youth. It emerged at a very specific time in human history and is shaped by the dynamic and potentially destabilizing traditions of modern capitalism. That explains why in many forms it has lacked the deep sensitivity to the biosphere that is present in the origin stories of indigenous peoples around the world.

The universe of the modern origin story is restless, dynamic, evolving, and huge. The geologist Walter Alvarez reminds us how big it is by asking how many stars it contains. Most galaxies have something like 100 billion stars, and there are at least that many galaxies in the universe. That means that there are (deep breath) 10,000,000,000,000,000,000,000 (10^{22}) stars in the universe.[4] New observations in late 2016 hinted that there may be many more galaxies in the universe, so feel free to add a few more zeros to this number. Our sun is a pretty ordinary member of that huge gang.

The modern origin story is still under construction. New sections are being added, existing parts still have to be tested or tidied up, and scaffolding and clutter need to be removed. And there are still holes in the story, so, like all origin stories, it will never lose a sense of mystery and awe. But in the past few decades, our understanding of the universe we live in has become much richer, and that may even enhance our sense of its mystery because, as the French philosopher Blaise Pascal wrote: "Knowledge is like a sphere; the greater its volume, the larger its contact with the unknown."[5] With all its imperfections and uncertainties, this is a story we need to know, just as the Lake Mungo people needed to know their origin stories. The modern origin story tells of the heritage all humans share, and so it can prepare us for the huge challenges and opportunities that all of us face at this pivotal moment in the history of planet Earth.

At the heart of the modern origin story is the idea of increasing complexity. How did our universe appear, and how did it generate the rich cavalcade of things, forces, and beings of which we are a part? We don't really know what it came out of or if anything existed before the universe. But we do know that when our universe emerged from a vast foam of energy, it was extremely simple. And simplicity is still its default condition. After all, most of our universe is cold, dark, empty space. Nevertheless, in special and unusual environments such as on our planet, there existed perfect Goldilocks conditions, environments, like Baby Bear's porridge in the story of Goldilocks, that were not too hot and not too cold, not too thick and not too thin, but just right for the evolution of complexity.[6] In these Goldilocks environments, increasingly complex things have appeared over many billions of years, things with more moving parts and more intricate internal relations. We should not make the mistake of assuming that complex things are necessarily better than simple things. But complexity does matter to us humans, because we are very complex, and the dynamic global society we live in today is one of the most extraordinarily complex things we know. So understanding how complex things emerged and what Goldilocks conditions allowed them to emerge is a great way of understanding ourselves and the world we live in today.

More complex things appeared at key transition points, and I will refer to the most important of these as *thresholds*. The thresholds give shape to the complicated narrative of the modern origin story. They highlight major turning points, when already existing things were rearranged or otherwise altered to create something with new, "emergent" properties, qualities that had never existed before. The early universe had no stars, no planets, and no living organisms. Then, step by step, entirely new things began to appear. Stars were forged from atoms of hydrogen and helium, new chemical elements were created inside dying stars, planets and moons formed from blobs of ice and dust using these new chemical elements, and the first living cells evolved in the rich chemical environments of rocky planets. We humans are very much part of this story, because we are products of the evolution and diversification of life on planet Earth, but in the course of our brief but remarkable history, we have created so many entirely new forms of complexity that, today, we seem to dominate change on our world. The appearance of something new and more complex than what preceded it, something with new emergent properties, always seems as miraculous as the birth of a baby, because the general tendency of the universe is to get less complex and more disorderly. Eventually, that tendency toward increasing disorder (what scientists term *entropy*) will win out, and the universe will turn into a sort of random mess without pattern or structure. But that's a long, long way in the future.

Meanwhile, we seem to live in a vigorous young universe that is full of creativity. The birth of the universe—our first threshold—is as miraculous as any of the other thresholds in our modern origin story.

Timeline

This timeline gives some fundamental dates for the modern origin story using both approximate absolute dates and recalculated dates, as if the universe had been created 13.8 years ago instead of 13.8 billion years ago. This second approach makes it easier to get a sense of the chronological shape of the story. After all, natural selection did not design our minds to cope with millions or billions of years, so this shorter chronology should be easier to grasp.

Most of the dates given for events that happened more than a few thousand years ago were established only in the past fifty years using modern chronometric technologies, of which the most important is radiometric dating.

EVENT	APPROXIMATE ABSOLUTE DATE	DATE DIVIDED BY 1 BILLION
THRESHOLD 1: Big bang: origin of our universe	13.8 billion years ago	13 years, 8 months ago
THRESHOLD 2: The first stars begin to glow	13.2 (?) billion years ago	13 years, 2 months ago
THRESHOLD 3: New elements forged in dying large stars	Continuously from threshold 2 to the present day	Continuously from threshold 2 to the present day
THRESHOLD 4: Our sun and solar system form	4.5 billion years ago	4 years, 6 months ago

EVENT	APPROXIMATE ABSOLUTE DATE	DATE DIVIDED BY 1 BILLION
THRESHOLD 5: Earliest life on Earth	3.8 billion years ago	3 years, 9 months ago
The first large organisms on Earth	600 million years ago	7 months ago
An asteroid wipes out the dinosaurs	65 million years ago	24 days ago
The hominin lineage splits from the chimp lineage	7 million years ago	2.5 days ago
Homo erectus	2 million years ago	17 hours ago
THRESHOLD 6: First evidence of our species, *Homo sapiens*	200,000 years ago	100 minutes ago
THRESHOLD 7: End of last ice age, beginning of Holocene, earliest signs of farming	10,000 years ago	5 minutes ago
First evidence of cities, states, agrarian civilizations	5,000 years ago	2.5 minutes ago
Roman and Han Empires flourish	2,000 years ago	1 minute ago
World zones begin to be linked together	500 years ago	15 seconds ago
THRESHOLD 8: Fossil-fuels revolution begins	200 years ago	6 seconds ago
The Great Acceleration; humans land on the moon	50 years ago	1.5 seconds ago
THRESHOLD 9 (?): A sustainable world order?	100 years in the future?	3 seconds to go
The sun dies	4.5 billion years in the future	4 years, 6 months to go
The universe fades to darkness; entropy wins	Gazillions and gazillions of years in the future	Billions and billions of years from now

PART I

Cosmos

CHAPTER 1

In the Beginning: Threshold 1

To make an apple pie from scratch, you must first invent the universe.

— CARL SAGAN, *COSMOS*

So it must have been after the birth of the simple light
In the first, spinning place, the spellbound horses
  walking warm
    Out of the whinnying green stable
      On to the fields of praise.

— DYLAN THOMAS, "FERN HILL"

Jump-Starting an Origin Story

Bootstrapping is the impossible task of lifting yourself into the air by pulling *really, really* hard on your bootstraps. The idea entered computer jargon (*booting* or *rebooting*) to describe how computers wake up from the dead and then load instructions telling them what to do next. Literally, of course, bootstrapping is impossible, because to lift something, you need something to provide leverage. "Give me a lever and a place to stand on," said the Greek philosopher Archimedes, "and I will move the Earth." But what could possibly leverage the creation of a new universe? How do you bootstrap a universe? Or, for that matter, the origin story that describes how a new universe appeared?

17

Bootstrapping origin stories is almost as hard as bootstrapping universes. One possible approach is to vanish the problem of beginnings by assuming the universe was always there. No bootstrapping needed. Many origin stories have gone this way. So have many modern astronomers, including those who supported the steady-state theory in the middle of the twentieth century. This is the idea that at large scales, the universe has always been pretty much as it is today. Similar, but subtly different, is the idea that, yes, there was a moment of creation when great forces or beings roamed the universe making things, but since then, nothing much has changed. The elders of Lake Mungo might have seen the universe like this, describing a world brought to life more or less in its current form by their ancestors. Isaac Newton saw God as the "first cause" of everything and argued that He was present in all of space. That is why Newton thought that the universe as a whole did not change much. The universe, he once wrote, was "the Sensorium of a Being incorporeal, living, and intelligent."[1] Early in the twentieth century, Einstein was so sure the universe was unchanging (at large scales) that he added a special constant to his theory of relativity to make it predict a stable universe.

Is the idea of an eternal or unchanging universe satisfying? Not really, particularly if you have to smuggle in a creator to kick-start the process, as in "In the beginning there was nothing, then God made..." The logical glitch is obvious, though it has taken some sophisticated minds a long time to see it clearly. At the age of eighteen, Bertrand Russell gave up on the idea of a creator god after reading the following passage in the autobiography of John Stuart Mill: "My father taught me that the question, 'Who made me?' cannot be answered, since it immediately suggests the further question, 'Who made God?'"[2]

And there's another puzzle. If a god is powerful enough to design a universe, that god must surely be more complex than the universe, so assuming a creator god means explaining a fantastically complex universe by imagining something even more complex that just...created it. Some might think that was cheating.

The ancient Indian hymns known as the Vedas hedge their bets. "There was neither non-existence nor existence then; there was neither the realm of space nor the sky which is beyond."[3] Perhaps everything arose from a sort of primordial tension between being and nonbeing, a murky realm that was not quite something but could *become* something. Perhaps, as a modern Australian Aboriginal saying puts it, nothing is *entirely* nothing.[4] It's a tricky idea, and some might dismiss it as fuzzy and mystical if it didn't have striking parallels to the modern idea, embedded in quantum physics, that space is never *entirely* empty but is full of possibilities.

Is there a sort of ocean of energy or potential from which particular forms emerge like waves or tsunamis? This is such a common concept that it is tempting to think our ideas about ultimate beginnings come from our own experiences. Every morning, we each experience how a conscious world, with shapes, sensations, and structures, seems to emerge from a chaotic unconscious world. Joseph Campbell writes: "As the consciousness of the individual rests on a sea of night into which it descends in slumber and out of which it mysteriously wakes, so, in the imagery of myth, the universe is precipitated out of, and reposes upon, a timelessness back into which it again dissolves."[5]

But perhaps this is too metaphysical. Maybe the difficulty is logical. Stephen Hawking argues that the question of beginnings is just badly put. If the geometry of space-time is spherical, like the surface of Earth but in more dimensions, then asking what existed before the universe is like looking for a starting point on the surface of a tennis ball. That's not how it works. There is no edge or beginning to time, just as there is no edge to the surface of Earth.[6]

Today, some cosmologists are attracted to another set of

concepts that tug us back to the idea of a universe without a beginning or end. Perhaps our universe is part of an infinite multiverse in which new universes keep popping out of big bangs. This could be right, but at present we have no hard evidence for anything before our own, local big bang. It's as if the creation of our universe was so violent that any information about what it came out of was erased. If there are other cosmological villages, we can't yet see them.

Frankly, today we have no better answers to the problem of ultimate beginnings than any earlier human society had. Bootstrapping a universe still looks like a logical and metaphysical paradox. We don't know what Goldilocks conditions allowed a universe to emerge, and we still can't explain it any better than novelist Terry Pratchett did when he wrote, "The current state of knowledge can be summarized thus: In the beginning, there was nothing, which exploded."[7]

Threshold 1: Quantum Bootstrapping a Universe

The bootstrap for today's most widely accepted account of ultimate origins is the idea of a big bang. This is one of the major paradigms of modern science, like natural selection in biology or plate tectonics in geology.[8]

It wasn't until the early 1960s that the crucial pieces of the big bang story emerged. That's when astronomers first detected the cosmic microwave background radiation (CMBR) — energy left over from the big bang and present everywhere in today's universe. Though cosmologists still struggle to understand the moment when our universe appeared, they can tell a rollicking story that begins about (deep breath, and I hope I've got this precise) a billionth of a billionth of a billionth of a billionth of a billionth of a second after the universe appeared (around 10^{-43} of a second after time zero).

The bare-bones story goes like this: Our universe began as a point smaller than an atom. How small is that? Our species' minds evolved to deal with things at human scales, so they struggle with things this tiny, but it might help to know that you could squeeze a million atoms into the dot at the end of this sentence.[9] At the moment of the big bang, the entire universe was smaller than an atom. Packed into it was all the energy and matter present in today's universe. All of it. That is a daunting idea, and at first it might appear plain crazy. But all the evidence we have at present tells us that this strange, tiny, and fantastically hot object really existed about 13.82 billion years ago.

We don't yet understand how and why this thing appeared. But quantum physics tells us, and particle accelerators—which speed up subatomic particles to high velocities by means of electric or electromagnetic fields—*show* us, that something really can appear in a vacuum from nothing, though grasping what this means requires a sophisticated understanding of *nothing*. In modern quantum physics, it is impossible to determine precisely the position and motion of subatomic particles. This means you can never say for sure that a particular region of space is empty, and that means that emptiness is tense with the possibility that something might appear. Like the "neither non-existence nor existence" of the Indian Vedas, this tension seems to have bootstrapped our universe.[10]

Today, we refer to the universe's first moment as the "big bang," rather as if, like a newborn baby, the universe yelled out at its birth. This cute term was coined in 1949 by an English astronomer, Fred Hoyle, who thought the idea was ridiculous. In the early 1930s, when the concept of a big bang was first floated, the Belgian astronomer (and Catholic priest) Georges Lemaître called the newborn universe the "cosmic egg" or the "primordial atom." It was clear to the few scientists who took the idea seriously that, with so much energy squashed up inside it, the primordial atom had to be inconceivably hot and had to be expanding like crazy to relieve the pressure. The expansion continues today; it's as if a vast spring has been uncoiling for more than thirteen billion years.

A lot happened in the first seconds and minutes after the big bang. Most important of all, the first interesting structures and patterns appeared, the first entities or energies that had distinctive *non*random forms and properties. The emergence of something with distinctive new qualities is always magical. We will see this happening over and over again in the modern origin story, although what appears to be magical at first may seem less so once we understand that the new thing and its new qualities did not arrive out of nowhere or from nothing. New things with new properties emerge from already existing things and forces that are arranged in new ways. It's the new arrangements that yield the new properties, just as arranging tiles in a different way can generate a new pattern in a mosaic. Take an example from chemistry. We normally think of hydrogen and oxygen as colorless gases. But join two hydrogen atoms to a single oxygen atom in a particular configuration, and you get a molecule of water. Put lots of those molecules together, and you get the utterly new quality that we think of as "wateriness." When we see a new form or structure with new qualities, we are really seeing new arrangements of what already existed. Innovation is emergence. If we think of emergence as a character in our story, it's probably slinky, mysterious, and unpredictable, likely to pop up from the darkness unexpectedly and take the plot in new and surprising directions.

The first structures and patterns in the universe emerged in just this way, as things and forces that popped out of the big bang were arranged in new configurations.

At the earliest moment for which we have some evidence, a split second after the big bang, the universe consisted of pure, random, undifferentiated, shapeless energy. We can think of energy as the *potential for something to happen,* the capacity to *do* things or *change* things. The energies inside the primeval atom were staggering, many trillions of degrees above absolute zero. There was a brief period of super-rapid expansion known as *inflation*. Expansion was so fast that much of the universe may have been projected far beyond anything we will ever see. That means that what we see today is probably just a tiny part of our entire universe.

A split second later, rates of expansion slowed. The turbulent energies of the big bang settled down, and as the universe kept expanding, the energies were spread out and diluted. Average temperatures fell, and they have kept falling, so today, most of the universe is just 2.76 degrees Celsius above absolute zero. (Absolute zero is the temperature at which nothing even jiggles.) We don't feel the chill, nor do any of the other organisms on planet Earth, because we are warmed by the campfire of our sun.

In the extreme temperatures of the big bang, almost anything was possible. But as temperatures dropped, possibilities narrowed. Distinct entities began to emerge like ghosts within the chaotic fog of the cooling universe, entities that could not exist in the violent cauldron of the big bang itself. Scientists call these changes of form and structure *phase changes*. We see phase changes in our daily lives when steam loses energy and turns into water (whose molecules move about a lot less than steam molecules) and when water turns into ice (which has so little energy that its molecules just jiggle in place). Water and ice can exist only in a narrow range of very low temperatures.

Within a billionth of a billionth of a billionth of a billionth of a second after the big bang, energy itself underwent a phase change. It split into four very different species. Today, we know them as gravity, the electromagnetic force, and the strong and weak nuclear forces. We need to get acquainted with their different personalities, because they shaped our universe. Gravity is weak, but it reaches across vast distances and always pulls things together, so its power accumulates. It tends to make the universe more clumpy. Electromagnetic energy comes in negative and positive forms, so it often cancels itself out. Gravity, though puny, shapes the universe on a large scale. But electromagnetism dominates at the level of chemistry and biology, so it's what holds our bodies together. The third and fourth fundamental forces are known, unexcitingly, as the strong and weak nuclear forces. They reach over tiny distances, so they matter on a subatomic scale. We humans don't experience them directly, but they shape every aspect of our world because they determine what happens deep inside atoms.

There may be other species of energy. In the 1990s, new measures of the universe's rate of expansion showed that the rate is increasing. Borrowing an idea first floated by Einstein, many physicists and astronomers now argue that there may be a form of antigravity that is present in all of space, so its power increases as the universe expands. Today, the mass of this energy may account for as much as 70 percent of the total mass of the universe. But even if it is beginning to dominate our universe, we don't yet understand what this energy is or how it works, so physicists call it *dark energy*. The term is a placeholder. Watch this space, because understanding dark energy is one of the great challenges of contemporary science.

Matter appeared within the first second after the big bang. Matter is the stuff that energy pushes around. Until just over a century ago, scientists and philosophers assumed that matter and energy were distinct. We now know that matter is really a highly compressed form of energy. The young Albert Einstein demonstrated this in a famous paper in 1905. That formula— energy (E) is equal to mass (m) times the speed of light (c) squared, or $E = mc^2$—tells us how much energy is compressed inside a given amount of matter. To figure out how much energy is locked up in a bit of matter, multiply the mass of the matter not by the speed of light (which is more than one billion kilometers per hour) but by the speed of light *times itself*. This is a colossal number, so if you uncompress a tiny bit of matter, you get a huge amount of energy. That's what happens when an H-bomb explodes. In the early universe, the opposite process occurred. Huge amounts of energy were compressed into tiny amounts of matter, like motes of dust in a vast fog of energy. Remarkably, we humans have managed to re-create such energies briefly, in the Large Hadron Collider outside Geneva. And, yes, particles do start popping out of that boiling ocean of energy.

And we're still in the first second...

The First Structures

Within the chaotic fog of energy just after the big bang, distinct forms and structures began to appear. Though the fog of energy is always there, the structures that emerged from it will give our origin story shape and a plotline. Some structures or patterns will last for billions of years, some for a split second, but *none* are conserved. They are evanescent, like waves on the ocean's surface. The first law of thermodynamics tells us that the ocean of energy is always there; it's conserved. The second law of thermodynamics tells us that all the forms that emerge will eventually dissolve back into the ocean of energy. The forms, like the movements of a dance, are *not* conserved.

Some distinct structures and forms emerged within a second of the big bang. Why? Why is the universe not just a random flux of energy? This is a fundamental question.

If our story had a creator god, explaining structure would be easy. We could just assume (as many origin stories do) that God preferred structure to chaos. But most versions of the modern origin story no longer accept the idea of a creator god because modern science can find no direct evidence for a god. Many people have *experiences* of gods, but those reported experiences are diverse and contradictory, and they cannot be reproduced. They are too malleable, too diffuse, and too subjective to provide objective, scientific evidence.

So the modern origin story has to find other ways of explaining the emergence of structures and forms. And that's not easy, because the second law of thermodynamics tells us that sooner or later, all structures will eventually break down. As the Austrian physicist Erwin Schrödinger wrote: "We now recognize this fundamental law of physics to be just the natural tendency of things to approach the chaotic state (the same tendency that the books of a library or the piles of papers and manuscripts on a writing desk display) unless we obviate it."[11]

If there is a bad guy in the modern origin story, it is surely entropy, the apparently universal tendency for structures to dissolve into randomness. Entropy is the loyal servant of the second law of thermodynamics. So, if we think of entropy as a character in our story, we should imagine it as dissolute, lurking, careless of others' pain and suffering, not interested in looking you in the eye. Entropy is also very, very dangerous, and in the end it will get us all. Entropy stands at the finale of all origin stories. It will dissolve away all structures, all shapes, every star and every galaxy and every living cell. Joseph Campbell described entropy's role poetically in a book on mythology: "The world, as we know it…yields but one ending: death, disintegration, dismemberment, and the crucifixion of our heart with the passing of the forms that we have loved."[12]

Modern science explains entropy's role in the cold-blooded language of statistics. Of all the myriad ways in which things can be arranged, the overwhelming majority are unstructured, random, disordered. Most change is like taking a pack of 10^{80} cards (that's 10 followed by 80 zeros, or roughly the number of atoms in the universe) and shuffling it again and again in the hope of finding all the aces next to each other. That's an inconceivably rare pattern, so rare that you are unlikely to see it even if you keep shuffling for many times the age of the universe. Most of the time you're going to find little or no structure. If you throw a bomb into a construction site full of bricks, mortar, wires, and paint, what are the odds that when the dust clears, you'll find an apartment building all wired up, decorated, and ready for buyers? The world of magic can ignore entropy, but our world can't. That's why most of the universe, particularly the vast empty spaces between galaxies, lacks shape and structure.

So powerful is entropy that it is not easy to understand how any structures appeared in the first place. But we know that they did. And they seem to have appeared with entropy's permission. It's as if, in return for letting things link up to form more complex structures, entropy demanded a complexity tax, to be paid in energy. In fact, we'll see that entropy has demanded many different types of complexity taxes, a bit like the Russian emperor Peter the Great, who formed a special government office to dream up new taxes. Entropy likes this deal because the taxes paid by all complex entities will help entropy's grim task of turning the entire universe into mush. The very act of paying entropy taxes creates more chaos and more waste, just as running a modern city generates huge amounts of garbage and heat. We all pay entropy's taxes every second of our lives. We will stop paying on the day we die.

So how did the very first structures emerge? This is a problem for which science does not yet have complete answers, though there are many promising ideas.

In addition to energy and matter, some basic operating rules emerged from the big bang. Scientists did not begin to understand how fundamental these rules were until the scientific revolution in the seventeenth century. Today, we describe these rules as the fundamental laws of physics. They explain why the frantic and chaotic energies of the primeval atom were not completely directionless—the laws of physics steered change down particular pathways and blocked a nearly infinite range of other possibilities. The laws of physics filtered out those states of the universe that were not compatible with them, so at any given moment, the universe existed in only one of the many states that *were* compatible with the universe's operating rules. These new states, in turn, generated more rules that steered change down new pathways.

This constant filtering out of impossible states guaranteed a minimum of structure. We don't know why the rules emerged or why they took the forms they did. We don't even know if these rules were inevitable. Perhaps other universes exist with slightly different rules. Perhaps in some universes, gravity is stronger or electromagnetism is weaker. If so, these universes' inhabitants (if they have any) will tell different origin stories. Maybe some universes lasted for a millionth of a second, while others will exist much longer than ours. Perhaps some universes generate many exotic life-forms while others are biological graveyards. If indeed our universe exists in a multiverse, then we can imagine a grand throwing of the dice when our universe was created, followed by an announcement: "Okay, there will be gravity in this universe, and electromagnetism as well, and electromagnetism is going to be 10^{36} times as strong as gravity." (That really is the ratio of the strength of gravity and electromagnetism, at least in our universe.) The existence of these rules ensured that our universe would never be totally chaotic. Something interesting was guaranteed to emerge somewhere.

There were structures and patterns as soon as energy emerged in distinct forms. When energy congealed into the first particles of matter, these, too, had rules. Neutrons, protons, and electrons, the basic constituents of atoms, appeared within seconds of the big bang, as did proton and electron antiparticles (that is, negatively charged protons and positively charged electrons), forming what physicists call *matter* and *antimatter*. As the universe plunged below the temperatures at which matter and antimatter could easily be created, there took place a violent, universe-wide demolition derby in which matter and antimatter annihilated each other, unlocking huge amounts of energy. Luckily for us, a tiny surplus of matter (perhaps one particle in a billion) survived the carnage. The leftover particles of matter got locked into place because temperatures were soon too low to turn them back into pure energy. And that leftover stuff is what our universe is made of.

As temperatures fell, matter diversified. Electrons and neutrinos were ruled by electromagnetism and the weak nuclear force. The protons and neutrons that form atomic nuclei were made from triplets of strange particles known as quarks, bound together by the strong nuclear force. Electrons, neutrons, quarks, protons, neutrinos... within just a few seconds of the big bang, our rapidly cooling universe had locked in some distinct structures, each with its own emergent properties. But as the hurricane of the big bang abated, the extreme energies needed to unlock these primordial structures vanished, and that's why, to us, the different forms of energy and particles such as protons and electrons seem more or less immortal.

This is how chance and necessity combined to produce the first simple structures. The laws of physics had filtered out many possibilities—that was the necessity part. Chance then rearranged things randomly from the possibilities that remained. This is how it all works. As nanophysicist Peter Hoffmann writes: "Tempered by physical law, which adds a dash of necessity, chance becomes the creative force, the mover and shaker of our universe. All the beauty we see around us, from galaxies to sunflowers, is the result of this creative collaboration between chaos and necessity."[13]

The First Atoms

Within a few minutes of the big bang, as protons and neutrons teamed up, more structures appeared. A single proton is the nucleus of a hydrogen atom; a pair of protons (with two neutrons) form the nucleus of a helium atom, so the universe was beginning to build the first atoms. But it takes a lot of energy to fuse protons because their positive charges repel each other, and temperatures were falling fast just after the big bang, so it was impossible to fuse more protons to form the nuclei of larger atoms. This explains a fundamental aspect of our universe: almost three-quarters of all the atoms in it are hydrogen, and most of the rest are helium.

A lot more matter consists of *dark matter,* stuff we don't yet understand, though we know it exists because its gravitational pull determines the structure and distribution of galaxies. So, a few minutes after the big bang, our universe consisted of vast clouds of dark matter in which were embedded crackling plasmas of protons and electrons with photons of light flowing through them. Today, we find plasmas only in the centers of stars.

Now we must pause and wait about 380,000 years (almost twice as long as our species has existed on Earth). During this time, the universe kept cooling. When temperatures fell below ten thousand degrees Celsius, there was one more phase change, like steam turning into water. To explain this phase change, we need to understand that heat is really a measure of the motion of atoms. All particles of matter are constantly jiggling about with energy, like nervous children, and temperature is a measure of the average degree of jiggling. The jiggling is real. In a famous paper published in 1905, Einstein showed that the jiggling of atoms causes the random gyrations of dust particles in the air. As temperatures drop, particles jiggle less, until eventually they can link up. As the universe cooled, the electromagnetic force tugged negatively charged electrons toward positively charged protons until the electrons calmed down enough to fall into orbits around protons. And voilà! We had the first atoms, the basic constituents of all the matter around us.

Normally, isolated atoms are electrically neutral, because the positive and negative charges of their protons and electrons cancel each other out. So when the first atoms of hydrogen and helium formed, most of the matter in the universe suddenly went neutral, and the tingling plasma evaporated. Photons, the carriers of the electromagnetic force, could now flow freely through an electrically neutral mist of atoms and dark matter. Today, astronomers can detect the results of this phase change, because photons that escaped the plasma generated a thin background hum of energy (the cosmic microwave background radiation) that still pervades the entire universe.

Our origin story has crossed its first threshold. We have a universe. Already it has some structures with distinctive emergent properties. It has distinct forms of energy and matter, each with its own personality. It has atoms. And it has its own operating rules.

What's the Evidence?

Bizarre as this story may seem when you hear it for the first time, we have to take it seriously, because it is supported by vast amounts of evidence.

The first clue that the big bang really happened was the discovery that the universe is expanding. If it's expanding now, logic tells us that at some time in the remote past, it must have been infinitesimally small. We know the universe is expanding because we have instruments and observational techniques that were not available to the people of Lake Mungo, even though we can be sure they were superb naked-eye astronomers.

Most astronomers since Newton's time assumed that the universe must be infinite, because if it was not, the laws of gravity should have gathered its contents into a single gluggy mass, like oil in a sump. By the nineteenth century, astronomers had instruments precise enough to start mapping the distribution of stars and galaxies, and the astronomical maps they created began to hint at a very different picture of the universe.

The mapping began with nebulae, fuzzy blurs that popped up on all their star charts. (We now know that most nebulae are entire galaxies, each with billions of stars.) How far away were the nebulae? What exactly were they? Were they moving? Over time, astronomers have learned how to tease out more and more information about stars from the light they emit. That information includes their distance from us and whether they are heading closer or moving away.

One of the cleverest methods to study the movement of stars and nebulae uses the Doppler effect (named after the nineteenth-century Austrian mathematician Christian Andreas Doppler) to measure the speed at which stars or nebulae are moving toward or away from us. Energy travels in waves, and waves, like those at the beach, have a frequency. They reach peaks at a regular pace that you can measure. But the frequency changes if you move. If you get in the ocean and swim out, the frequency at which you encounter waves will seem to increase. The same thing happens with sound waves. If an object, such as a motorbike, is making a noise and moving toward you, the frequency of the sound waves will seem to increase, and your ears will interpret the higher frequency as a higher pitch. After it passes you, the pitch will seem to drop, because now the waves are being stretched out. The rider, of course, is not moving relative to the motorbike and keeps hearing the same pitch. The Doppler effect is the apparent change in frequency of electromagnetic emissions as objects move toward or away from each other.

The same principle works with starlight. If a star or galaxy is moving toward Earth, the frequency of its light waves will seem to increase. Our eyes interpret higher-frequency visible light as blue light, so we say it has shifted toward the blue end of the electromagnetic spectrum. But if it is moving away from Earth, the frequency of its light will seem to shift toward the red end of the spectrum; astronomers say it is redshifted. And we can tell how fast a star or galaxy is moving by measuring how much the frequency has shifted.

In 1814, a young German scientist, Joseph von Fraunhofer, created the first scientific spectroscope, a specialized prism that splits up the frequencies of starlight just as a glass prism splits light into the colors of the rainbow. Fraunhofer found that spectra from sunlight had thin dark lines at particular frequencies, like cosmological bar codes. Two other German scientists, Gustav Kirchhoff and Robert Bunsen, eventually showed in the lab that particular elements emit or absorb light energy at specific frequencies. It seemed that the dark lines were the result of light from the sun's core being absorbed by atoms of different elements in the sun's cooler outer regions. This reduced the energy at those frequencies, leaving dark lines on the emission spectrum. We call these dark lines *absorption lines,* and different elements generate different patterns of absorption lines. For example, there are lines that are typical of carbon and iron. If starlight is redshifted, then all these lines shift to the red end of the spectrum, and we can even measure exactly how far they have shifted. This is the astronomer's equivalent of a police speed trap.

In the early twentieth century, an American astronomer, Vesto Slipher, used these techniques to show that a surprising number of astronomical objects were redshifted—that is, they were moving away from Earth, and quite fast. That scattering was very strange. Its real meaning became clear only when another American astronomer, Edwin Hubble, combined these findings with measurements of the distance to these remote objects.

Estimating the distance to stars and nebulae is tricky. In principle, as the Greeks understood, you could use the parallax method, like a surveyor. Over the months, as Earth swings around the sun, watch to see if some stars in the night sky seem to move relative to other stars. If they do, you can use trigonometry to figure out how far away they are. Unfortunately, even the nearest star, Proxima Centauri, is so distant (about four light-years from Earth) that you cannot detect any motion without fancy equipment. Not until the nineteenth century were astronomers able to measure the distance to nearby stars using parallax. But in any case, the objects Vesto Slipher was studying were much more distant.

Fortunately, in the early twentieth century, Henrietta Leavitt, a Harvard Observatory astronomer, found a way to measure the distance to remote stars and nebulae using a particular type of star known as a Cepheid variable, a star whose brightness varies with great regularity (the polestar is a Cepheid). She found a simple correlation between the frequency of the variations and the star's luminosity, or brightness, so she could calculate a Cepheid's absolute brightness. Then, by comparing that with the apparent brightness the star had when seen from Earth, she could calculate how far away it was, because the amount of light from a star diminishes by the square of the distance through which it travels. This wonderful technique provided the astronomical standard candles that Edwin Hubble needed to make two profound discoveries about our universe.

Early in the twentieth century, most astronomers believed that the entire universe was contained within our galaxy, the Milky Way. In 1923, Hubble used one of the world's most powerful telescopes, at the Mount Wilson Observatory in Los Angeles, to show that Cepheid variables in what was then known as the Andromeda nebula were so far away that they could not be in our own galaxy. This proved what some astronomers had suspected: that the universe was much larger than the Milky Way and consisted of many galaxies, not just our own.

Hubble made an even more astonishing discovery as he began to measure the distance to large numbers of distant objects using Cepheid variables. In 1929, he demonstrated that almost all galaxies appeared to be moving away from us and that the most remote objects seemed to have the largest red-shifts. In other words, the more distant an object was, the faster it was moving away. And *that* seemed to mean that the entire universe was expanding. The Belgian astronomer Georges Lemaî-tre had already suspected this on purely theoretical grounds. And, as Lemaître pointed out, if the universe was currently expanding, at some time in the past, everything in it must have been compressed into a tiny space, something he described as the *primordial atom.*

Most astronomers were shocked by the idea of an expanding universe and assumed there was an error in Hubble's calculations. Hubble himself was not at all sure about it, and Einstein was so convinced the universe was stable that he fiddled with the equations of general relativity so they would predict a stable universe, by adding what he called a *cosmological constant.*

Astronomers were skeptical partly because there really were problems with Hubble's estimates. He calculated that the expansion of the universe had begun just two billion years ago, yet astronomers already knew that Earth and its solar system were much older than that. That is one reason why, for several decades, most astronomers regarded the idea of an expanding universe as intriguing but probably wrong. Many preferred the alternative idea of a steady-state universe, proposed in 1948 by Hermann Bondi, Thomas Gold, and Fred Hoyle. Yes, agreed the steady-staters, galaxies seemed to be moving apart, but new matter was being created at the same time, so at large scales, the universe remained at about the same density and changed little.

Eventually, though, the evidence tipped in favor of an expanding universe. In the 1940s, Walter Baade, working at the Mount Wilson Observatory in LA (the same observatory at which Hubble had worked), showed there were two types of

Cepheid variable stars, and they yielded different estimates of distance. Baade's revised calculations suggested that the big bang might have happened more than 10 billion years ago (current best estimates suggest it occurred as much as 13.82 billion years ago). This eliminated the chronology problem. Today we know of no astronomical objects older than 13.82 billion years, which is a strong argument in favor of big bang cosmology. After all, if the universe were unchanging and eternal, there really should be lots of objects more than 13.8 billion years old.

The clinching evidence came in the mid-1960s, and it involved the discovery of cosmic microwave background radiation (CMBR). This is the radiation released when the first atoms formed, about 380,000 years after the big bang. The CMBR turned out to be the crucial proof of an expanding universe. Why?

By the 1940s, some astronomers and physicists were impressed enough by Hubble's data that they tried to figure out what might have happened if there really had been a big bang. What would the universe have been like at the start if everything was crushed into a primordial atom? If Hubble and Lemaître were right, the early universe would have been extremely dense and hot, and it must have been expanding and cooling fast. How would matter and energy behave under such extreme conditions? During the Second World War, the Manhattan Project to build an atomic bomb had encouraged research into the physics of very high temperatures. In the late 1940s, the Russian-born physicist George Gamow used insights from the Manhattan Project to figure out what had probably been going on in the universe just after the big bang. With a colleague, Ralph Alpher, he predicted that the universe would have eventually cooled enough for atoms to form, and when the first atoms formed, there should have been a huge release of energy as photons escaped the charged plasma of the preatomic era and began to flow freely through an electrically neutral universe. Further, they argued that this flash of energy should still be detectable, though its frequency would have fallen to near zero as it was stretched across an expanding universe. If scientists looked carefully enough, they would find radiation at temperatures close to absolute zero coming from all directions. To many this seemed a crazy idea, which was why no one started looking for low-temperature radiation pervading the entire universe.

In 1964, Gamow's flash of radiation was detected by accident. At Bell Labs in Holmdel, New Jersey, two radio astronomers, Arno Penzias and Robert Wilson, were building a high-precision radio antenna to communicate with artificial satellites. To eliminate interference, they cooled down the receiver to about 3.5 degrees Celsius above absolute zero, but there remained a puzzling hum of low-temperature energy. It seemed to come from all directions, so they knew it was not from some massive stellar explosion. Suspecting a glitch in their receiver, they removed a pair of pigeons roosting in the hornlike antenna and cleaned out the droppings, but it made no difference. (Sadly, the pigeons kept trying to return to the antenna and eventually had to be shot.) Nearby, in Princeton, a team of astronomers led by Robert Dicke had just started to look for Gamow's background radiation when they heard what Penzias and Wilson had found. They immediately realized they had been scooped. The two teams decided to collaborate on papers describing the discovery. They argued that it was probably the energy from just after the big bang that Gamow had predicted.

The discovery of cosmic microwave background radiation persuaded most astronomers that the big bang was real because no other theory could explain this all-pervading radiation. Making an odd but ultimately successful prediction like this is one of the most powerful ways of persuading scientists that your theory is sound. The universe, it seemed, really was expanding, and it really had been created in a big bang.

Today, the evidence that our universe began in a big bang is overwhelming. Lots of details remain to be worked out, but for the time being, the core idea is firmly established as the first chapter of a modern origin story. That's the bootstrap. And, as quantum physics allows things to appear from a vacuum, it seems that the entire universe really did pop out of a sort of nothingness full of potential.[14]

CHAPTER 2

Stars and Galaxies: Thresholds 2 and 3

Mankind is made of star stuff.
— HARLOW SHAPLEY, *VIEW FROM A DISTANT STAR*

The big bang gave us a universe, but for several hundred million years the universe was extremely simple. Beneath the surface, though, interesting new possibilities were stirring, and eventually, stars and galaxies began to light up the darkness. They added an entirely new cast of characters, new emergent properties, and new forms of complexity, and they led the universe across a second threshold of increasing complexity. To explain how these majestic new objects emerged, we need to go back to the beginning.

Free Energy: The Driver of Complexity

In the seconds and minutes after the big bang, the universe was in thermodynamic free fall. For a dazzling few moments, there was enough energy to make and unmake exotic new forms of energy and matter. But as temperatures plummeted, energy and matter froze into a few simple structures. In the kiln of the big bang, forces and particles stabilized like fired pottery. Together, the violent energies of the big bang and a few simple operating rules had created structures such as protons and electrons that would prove remarkably stable, because the temperatures that created them would rarely appear again in a cooling universe.

Then the rapid descent slowed, rather as if the universe were falling down a thermodynamic mountain into a valley. Gradients flattened, temperatures dropped less precipitously, and the pace of change decreased as the thermodynamic cliff face of the early universe gave way to a flatter, undulating landscape in which temperatures could rise as well as fall. Now it got harder to lock new structures in because they could be unraveled by even modest increases in heat. Atoms, for example, fell apart inside the first stars when temperatures rose above about ten thousand degrees Celsius.

In these less predictable environments, complex structures needed extra bracing if they were to stabilize. That bracing was provided by controlled, nonrandom flows of energy. Stars are held together by flows of energy generated in their cores. Living organisms, including you and me, are held together by delicate and precisely directed flows of energy managed by intricate metabolic processes in our cells. In a post–big bang universe, it takes work to build and maintain new complex structures. This is why there is a deep link between form, complexity, and directed or structured flows of energy.

Structured flows of energy is an intuitive description rather than a piece of scientific jargon. But here's the idea it's getting at: Thermodynamic theory distinguishes between energy flows that are completely random and energy flows that have direction, structure, and coherence so they can do work. Structured flows of energy are known as *free energy,* and unstructured flows are known as *heat energy.* The difference is not absolute. We're really talking about degrees of coherence or randomness. Nevertheless, the distinction between free energy and heat energy is fundamental to our origin story.

The first law of thermodynamics tells us that the total

amount of energy in the universe never changes. It is conserved. Our universe seems to have arrived with a fixed *potential for things to happen*. So the first law is really telling us about the primordial ocean of possibilities. The second law of thermodynamics tells us that the things that emerge from the ocean of possibilities can be more or less structured, like the ripples in a stream. But we should expect most of them to be less structured and become even less structured over time. That is because most possible arrangements of matter and energy have little or no structure, and if by chance you do find structure, expect it to decay fast.

A waterfall is a good illustration. Here, we have a lot of structure, but it will eventually dissipate. The water molecules at the top of the falls don't move about randomly, like molecules in a jar of air. They move in the same direction, like prowling cats, packing as close as they can. This is because, unlike gas molecules, which move as individuals, liquid molecules are held together by electromagnetism. So gravity can move them in close formation and in the same direction, like soldiers on the march. As water pours over the edge, potential energy turns into kinetic energy, or energy of motion. This is coordinated movement in a single direction. It's structured, so we can describe the energy that drives it as *free energy*. And free energy, unlike the random heat energy of gas molecules, can do work because it has some structure and shape and can push things in a single direction rather than pushing them every which way.[1] If you wanted to, you could direct this flow of free energy through a turbine and generate electricity. Free energy is what gets things done. It's the fast-moving, unstoppable Energizer bunny of our origin story.

But unlike energy in general, free energy is not conserved. It's unstable, like an uncoiling spring. As it does work, it loses both its structure and the ability to do more work. When the water from a waterfall smashes into rocks at the bottom, it turns into the scattered, incoherent energy of heat. Every molecule jiggles around more or less independently. The energy is still there; it's still conserved (that's the first law). But the molecules push in so many directions that their energy can no longer drive a turbine. Free energy has turned into heat energy. The second law of thermodynamics tells us that, in the very long run, all free energy will turn into heat energy.

Heat energy, like a drunken traffic cop, directs energy every which way and creates chaos. Free energy, like a sober traffic cop, directs energy down particular routes and creates order. Luckily for us, there was some free energy in the early universe because of our universe's basic operating rules. Those rules steered energy down particular, nonrandom pathways and ensured at least a minimum of structure.

Galaxies and Stars: Threshold 2

Free energy drove the emergence of the first large structures: galaxies and stars. The crucial source of free energy for this part of our origin story was gravity. Like a cosmological sheep-dog, gravity likes to herd things. And the things it herded were the simple forms of matter created in the big bang. Together, gravity and matter provided the Goldilocks conditions for the emergence of stars and galaxies.

Studies of the cosmic microwave background radiation show that in the early universe, there was little structure at large scales. Think of a gossamer-thin mist of hydrogen and helium atoms floating in a warm bath of dark matter permeated with photons of light. And all of it at more or less the same temperature. We know the early universe was homogenous because we can measure temperature differences in the CMBR, and we find that the hottest parts of the early universe were only about one one-hundredth of a degree warmer than the coolest parts. No usable temperature gradients here, no waterfalls of energy that could build new structures. You could generate a much larger temperature difference right now by rubbing your finger across your face.

Then gravity began to shape this unpromising material into something more interesting. While the big bang was pushing space apart, gravity was hustling around trying to pull energy and matter together.

The idea of gravity was central to Newton's understanding of the universe and provided one of the unifying ideas of the scientific revolution. Newton explained how gravity functions in one of the most important scientific works of all time: the *Philosophiae Naturalis Principia Mathematica,* or *The Mathematical Principles of Natural Philosophy,* published in 1687. Newton saw gravity as a universal force of attraction that operated between all masses. Two and a half centuries later, Einstein showed that energy could also exert a gravitational pull, because energy is what matter is made from.

Einstein made another important prediction about gravity: It was a form of energy, so, like electromagnetism or sound, it ought to generate waves. But Einstein feared the waves would be so tiny, no one would ever detect them. In September 2015, in a beautiful display of science at its best, gravity waves were finally detected by two huge machines, one in Louisiana and one in Washington State, operated by the Laser Interferometer Gravitational-Wave Observatory, or LIGO. In 2017, three of the men who contributed significantly to the project were awarded the Nobel Prize in Physics. The gravitational waves LIGO detected were generated about one hundred million years ago, when two black holes collided in a distant galaxy somewhere in the southern skies. (When they collided, dinosaurs still ruled our planet.) On Earth, each LIGO machine split beams of light in two and sent them traveling at right angles to each other up and down two four-kilometer tubes with mirrors at either end. When they returned after almost three hundred trips, they didn't arrive at exactly the same time. Tiny gravitational waves had stretched the tubes in one direction and shrunk them in the other by a distance much less than the width of a proton. Now that astronomers know that gravitational waves exist, they are hopeful they can use them to study the universe in new ways.

From the point of view of gravity, the early universe was too smooth. It needed to be clumped up. This tendency of gravity to rearrange the universe explains why we can think of the early universe as having low entropy, a sort of tidiness that entropy could mess up over the next few billion years. Once it got going, gravity took just a few hundred million years to turn the smooth particle mist of the early universe into a messier and lumpier space full of stars and galaxies.

As Newton showed, the strength of gravity increases where there is more mass and where things are closer together. That's why Earth exerts a much greater gravitational pull on objects than you do, and it's also why it tugs more gently on you if you are farther away from its surface—say, in the International Space Station. Now focus in on a small cube of the early universe's particle mist. Let's imagine that, quite randomly, the dark matter and atoms are slightly more concentrated in one corner of the cube than in another. Newton's laws tell us that gravity will be stronger in the denser corner, so here matter will get pulled together more forcefully, and the difference between denser and emptier regions will get magnified. In this way, cube by cube, gravity made the universe grainier and clumpier over millions of years.

As gravity forced atoms together, they collided more often and jiggled more frenetically. That drove up temperatures in the clumpier regions, as more heat was concentrated in smaller volumes of space. (The same principle explains why a tire gets warmer when you pump it up.) While most of the universe kept cooling, the clumpy bits began to heat up again. Eventually, some clumps got so hot that protons could no longer hold on to their electrons. Atoms broke apart, re-creating inside each clump the charged plasma, crackling with electricity, that had once pervaded the entire universe.

As gravity piled on the pressure, denser regions got denser, their cores got hotter, and gravity began to re-create the high energies of the early universe. At roughly ten million degrees Celsius, protons have so much energy that they can collide violently enough to overcome the repulsion of their positive charges. Once pushed across this barrier, protons began to link up in pairs, bound together by the strong nuclear force, which operates only over tiny distances. Proton pairs formed helium nuclei as they had done, briefly, once before, just after the big bang.

As protons fused, some of their mass was turned into pure energy, and, as we have seen, even a tiny bit of matter contains a colossal amount of energy. The same huge energies are released by H-bombs, which are powered, like every star, by fusion. So, as the core of a dense cloud of matter crosses the critical threshold of about ten million degrees, trillions of protons start fusing into helium nuclei, creating a furnace that releases colossal amounts of energy. Once lit, the furnace will keep burning as long as there are enough spare protons for fusion to continue.

The huge energies released by fusion will heat the core so that it expands and pushes back against gravity. Now the whole new structure will stabilize for millions or billions of years. A star has been born.

A Universe with Galaxies and Stars

But not just one star; in each clumpy region, there were billions of stars, and now the vast star cities we call galaxies began to glitter, lighting up the darkness of the young universe.

This universe with galaxies and stars is very different from the universe of the first atoms. Now the universe has structure at large scales as well as small, and we can say that the whole universe is more complex. There are dark, empty areas between galaxies, and bright, dense areas inside galaxies. Galaxies are thick with matter and energy, while the space between them is cold and empty. No longer smeared out like a mist, the interesting stuff is concentrated in vast sheets and filaments of galaxies, rather like the threads of a spider's web. Each galaxy has a particular structure. Most are spiral galaxies, like our home galaxy, the Milky Way, with hundreds of billions of stars revolving slowly around a dense core in which there is usually a black hole. But galaxies that collided with other galaxies often got messed up to form "irregular galaxies." Galaxies, in turn, were bound by gravity into clusters, and into clusters of clusters, creating stellar archipelagoes stretching across the entire universe.

Dotted through the universe, like hot raisins in a cold pudding, are individual stars that also have a lot of structure and new emergent properties. Each star has a hot core in which protons fuse together, generating energy that pushes back against gravity. Above the core, outer layers press down and supply it with proton fuel. The star's life history will depend primarily on its birth mass: how much stuff it contains at the start. Massive stars generate more gravitational pressure, so they are much hotter than stars with less mass. This explains why they burn their fuel fast and shut down within just a few million years. Stars with less mass burn more slowly, and many small stars will keep burning for much longer than the present age of the universe.

This more diverse universe had more varied environments, greater creative potential, and lots of energy gradients. There were gradients of light, temperature, and density, down which free energy flowed, like water over a waterfall. Each star poured energy into the cold spaces around it, generating flows of heat, light, and chemical energy that could be used to build new forms of complexity in nearby regions. Those are the flows of free energy that allow life to flourish here on planet Earth.

Gravity had kick-started the transformation of matter into stars by fusing protons despite the barrier created by their positive charges. This is a pattern we will see over and over again. It's a bit like the cup of coffee that helps you get going in the morning. Chemists refer to this initial shot of energy as *activation* energy; it's the energy of a lit match that starts a conflagration. One kind of energy changes something so as to release other flows of free energy that are much greater than the activation energy. In the story of star formation, gravity provided the activation energy for fusion and star formation and for all that followed.

But there's a puzzle here. What about the second law of thermodynamics? Entropy hates structure, so why does it allow more complex things to appear?

If you look closely at the energy flows, you'll see that complex structures, such as stars, pay dearly for their complexity. Look at all the energy from fusion. The first thing that energy does is prop up the star, preventing it from collapsing. This is a bit like a fee paid to entropy, a sort of complexity tax. When the star stops generating energy, it will collapse. The idea of a complexity tax helps explain an important phenomenon noted by the astrophysicist Eric Chaisson: roughly speaking, more complex phenomena need more dense flows of energy, more energy per gram per second. He estimates, for example, that the density of energy flowing through modern human society is about one million times greater than the density of energy flowing through the sun, while energy flowing through most living organisms lies somewhere between these extremes. It's as if entropy demands more energy from an entity if it tries to get more complex; more complex things have to find and manage larger and more elaborate flows of free energy. No wonder it's harder to make and maintain more complex things, and no wonder they usually break down faster than simpler things. This is an idea that runs right through the modern origin story and has a lot to tell us about modern human societies.[2]

Entropy loves this deal because the energy that props up a star, like the energy of a waterfall, eventually degrades when it is released into space. So, while the star is getting more complex, it's also helping entropy degrade free energy into heat energy. This is something we will see throughout the modern origin story. Increasing complexity is not a triumph over entropy. Paradoxically, the flows of energy that sustain complex things (including you and me) are helping entropy with its bleak task of slowly breaking down all forms of order and structure.

New Elements and Increasing Chemical Complexity: Threshold 3

A billion years after the big bang, the universe, like a young child, was already behaving in interesting ways. But chemically speaking, it was very boring. It contained just hydrogen and helium. Our third threshold of increasing complexity yielded new forms of matter: all the other elements of the periodic table. A universe with more than ninety distinct elements could do so much more than a universe with just hydrogen and helium.

Hydrogen and helium were the first elements to be made because they are the simplest. Hydrogen has one proton in its nucleus, so we say it has atomic number 1. Helium has two protons in its nucleus, so its atomic number is 2. When the CMBR was emitted, about 380,000 years after the big bang, there was also a sprinkling of lithium (atomic number 3) and beryllium (atomic number 4). And that was it. These were the only elements created in the big bang.

The Goldilocks conditions for creating more elements with larger nuclei were simple: lots of protons and very high temperatures, temperatures that had not existed since just after the big bang. Those temperatures would be created inside the dramatic, conflicted world of dying stars as they wearied, staggered, and eventually broke down, no longer able to pay entropy's complexity taxes.

To understand how stars manufacture new elements in their death throes, we need to understand how they live and age.

Stars live for millions or billions of years, so we cannot watch them aging. That's why the modern story of their life and death could not have been told by naked-eye astronomers such as the Maya or the people of Lake Mungo or ancient Athens. Our modern understanding is based on research from all over the world using instruments and data stores created only in the past two centuries. These allow modern astronomers to share information on millions of stars at different stages in their lives. As the English astronomer Arthur Eddington put it, astronomy is like walking through a forest with saplings, mature trees, and ancients close to death.[3] By studying trees at different points in their life cycles, you can eventually figure out how they grow, mature, and die.

For astronomers, there is one fundamental map that brings together a huge amount of information about stars: the Hertzsprung-Russell diagram. It's the astronomer's equivalent of the globes that used to sit in school classrooms, and, like those globes, it helps us make sense of a lot of information.

The Hertzsprung-Russell diagram, created circa 1910, classifies stars according to two basic properties. The first property, plotted on the vertical axis, is their intrinsic brightness or luminosity—that's really the amount of energy they send out into space—compared to the sun. The second property is their color, which tells you their surface temperature in kelvins (K). This is usually plotted on the horizontal axis. Because these two quantities change during a star's lifetime, the graph can help us understand the biographies of different types of stars. Major differences in the life histories of stars depend mainly on one more statistic: the mass of the cloud of matter from which they formed. High-mass stars have different biographies than low-mass stars.[4]

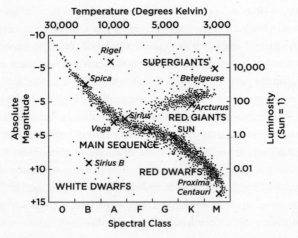

Hertzsprung-Russell diagram, simplified version with approximate positions of examples of different star types

On a Hertzsprung-Russell diagram, the most luminous stars, those emitting the most energy, such as Sirius, are toward the top. These are normally the stars with the most mass. The least luminous stars, such as our neighbor Proxima Centauri, are lower down. Our sun (at a luminosity of 1) is in the middle. Stars with very high surface temperatures are off to the left, and those with low surface temperatures are off to the right.

There are three main areas of interest in the diagram. Crossing the diagram, in a broad, curved band extending from the bottom right to the top left, is the main sequence. Most stars will spend about 90 percent of their lifetimes at some point on the main sequence. Where they sit depends on their mass, but all stars on the main sequence generate the energy they need by fusing protons into helium nuclei. And that's what our sun is doing right now, too. It is middle-aged and still on the main sequence. In the top right of the diagram you find red giants, like Betelgeuse, which is at one corner of the constellation Orion. These are aging stars that have used up most of the protons in their cores and are fueling their furnaces by burning other, larger nuclei. They have cooler surfaces because they have expanded to perhaps two hundred times the radius of our sun. But the total amount of light they emit is huge because they are very large, which is why they are near the top of the diagram. The third important region is in the bottom left-hand corner. Here, you find white dwarfs. These were red giants until they lost most of their outer layers, leaving just hot, dense cores.

When stars get very old, eventually they run out of free protons and their cores start filling up with an ash of fused protons—in other words, helium nuclei. Fusing helium nuclei requires much higher temperatures than fusing single protons, so eventually, the furnaces at their core stop working. When that happens, gravity takes over, and the stars collapse under their own mass. But that's not yet the end of the story. After a star collapses, it heats up again as gravity piles on the pressure. Far from the core, the star's outer layers expand and cool to keep everything in balance. To us, these cooler outer layers look red, which is why we call stars at this stage *red giants*. When our sun reaches this stage, it will expand to about two hundred times its present size and vaporize the inner planets, including Earth.

If the red giant has enough mass, gravity will compress it so tightly that its core gets hotter than ever before, hot enough to start fusing helium nuclei into heavier nuclei, such as carbon (with six protons) and oxygen (with eight protons). The star has revived, but fusing helium nuclei is a more complicated process than fusing protons and generates less energy, so stars at this stage have a much shorter life expectancy. Very large stars will go through several stages of increasingly frenetic expansion and contraction. Carbon and oxygen will fuse to form elements from magnesium to silicon and eventually iron. As the stars heat up, another mechanism kicks in, turning some neutrons into protons to create new types of nuclei. The core will gradually become a huge ball of iron surrounded by layers of other elements.

And that's the end of the road, because you cannot generate energy by fusing iron nuclei. Eventually, most stars will blast away their outer layers and end up as white dwarfs, which are down in the bottom left corner of the Hertzsprung-Russell diagram. White dwarfs are stellar zombies, with no furnace at their heart. They are extremely dense, often the size of Earth, but with the mass of the sun. If you try to lift a teaspoon of white-dwarf stuff, you'll fail, because it weighs at least a ton.[4] Though still hot, their corpses will cool over billions of years. But they have done their work by fertilizing their surroundings with new elements. Some white dwarfs die more spectacularly in vast supernova explosions if they get sucked into nearby stars. These explosions are so hot that they can create many of the elements in the periodic table. The spectacular death-by-explosions of white dwarfs generates what are known as type 1a supernovas. These all blow up at about the same temperature, so if you see one, you know how bright it is, and that means you can estimate its true distance. Type 1a supernovas allow astronomers to estimate distances hundreds of times farther away than Cepheid variables.

Stars more than about seven times the mass of our sun will also end their lives spectacularly in another type of explosion, known as a core-collapse supernova. When the core has formed a ball of iron larger than our sun, the central furnace will shut down for the last time. Gravity will smash the core together in a fraction of a second and with extreme violence, creating energies and temperatures higher than ever before in the star's lifetime. The star will explode in a supernova and may briefly emit as much energy as an entire galaxy. In just a few minutes, it manufactures many of the remaining elements of the periodic table and blasts them into space. Perhaps the most famous example of a core-collapse supernova lies at the heart of the Crab Nebula. Betelgeuse could go supernova at any time in the next million years.

Most supergiants, having blasted away their outer layers in supernovas, will contract so violently that protons and electrons are squashed together to form neutrons. Now the entire massive blob is crushed into a *neutron star,* an object made of neutrons packed together as closely as the particles in an atomic nucleus. This is a very unusual and extremely dense form of matter, as most atoms consist mainly of empty space. A neutron star just twenty kilometers across would weigh twice as much as our sun, and a teaspoon of neutron-star stuff would weigh a billion tons.[5] There is some evidence that many of the heavier elements in the periodic table may have been formed, not in standard supernovas, but during violent mergers of neutron stars.

Neutron stars spin rapidly, like warning beacons, and they were first detected in 1967 from a series of rapid flashes of energy. Rotating neutron stars are known as pulsars. Soon after the first pulsar was discovered, another one was detected in the heart of the Crab Nebula, the remnants of a supernova observed by Chinese astronomers in 1054. The Crab Nebula pulsar is about the size of a city and rotates thirty times every second.

For the most massive stars, there is another, even stranger, denouement. Their cores implode so violently that nothing can resist the collapse, and they turn into black holes, the densest objects we know of. Einstein predicted the existence of black holes, objects so dense that nothing can escape their gravitational pull, not even light, which is why we know so little about their innards. Black holes are very strange astronomical monsters, but we now have plenty of evidence that they are real. The first stars in our universe were probably huge, so it is likely that many collapsed into large black holes, and these may have provided the gravitational seeds around which whole galaxies formed, like pearls around grains of sand. Today, astronomers can detect large black holes at the centers of most galaxies, including our own. Their huge gravitational fields can suck nearby stars into their jaws. As a star is pulled across the border of a black hole (its event horizon), the star emits huge amounts of energy in a sort of death scream. These dying shrieks give rise to the exceptionally bright objects known as quasars.

The border, or event horizon, of a black hole is a point of no return. It represents a limit to our knowledge, because so little information can escape the clutch of a black hole. We can estimate the mass of the object that formed a black hole as well as its rotation. But that's more or less it. However, Stephen Hawking showed that subtle quantum effects allow tiny amounts of energy to leak out from black holes. Perhaps they are also leaking information, but if so, we don't yet know how to read it.

In these different ways, dying stars enriched and fertilized the young universe. Once forged in dying stars and supernovas, the elements of the periodic table gathered in huge dust clouds between stars; atoms combined to form simple molecules, and, by a sort of fermentation, they brewed up new forms of matter.

We know so much about stars because astronomers have developed techniques to determine what's going on inside stars millions of light-years from Earth. We have already seen how much information astronomers can tease from starlight. But visible light makes up only a tiny portion of the energies emitted by stars and galaxies. Modern telescopes let astronomers study emissions at all frequencies of the electromagnetic spectrum, from the longest and laziest of radio waves to the tiniest and most hyperactive gamma rays. Computers allow us to process enormous amounts of information with great precision, and space-based telescopes, such as the Hubble telescope, allow astronomers to observe the universe free of the distortions created by Earth's atmosphere. These modern scientific toys tell us a huge amount about our galactic environment.

Older instruments, such as optical telescopes and spectroscopes, have also been immensely important. The absorption lines revealed by spectroscopes tell us what elements are inside stars and in what proportions. Want to know how much gold there is in the sun? Point your spectroscope at the sun, study the absorption lines for gold, and measure how dark they are. You'll find out that gold makes up less than a trillionth of the mass of the sun. But the sun is so large that extracting all that gold would make you extremely rich, because it would yield much more gold than exists on planet Earth.

Astronomers can tell a star's surface temperature from the color (or frequency) of the light it emits, so we know that surface temperatures can be as low as 2,500 K and as high as 30,000 K. And, as we have already seen, they can calculate the total *amount* of light a star emits (its luminosity) by measuring its *apparent* brightness and then calculating how much brighter it would be close up. These two measurements—surface temperature and luminosity—provide the basic inputs for the Hertzsprung-Russell diagram. Finally, if we know a star's luminosity, we can often estimate its mass. Similar techniques help us estimate the distance, size, motion, and energy of entire galaxies.

These techniques have revolutionized our understanding of stars and galaxies in the past fifty years. They have helped us understand the evolution of stars and galaxies, how they break down, and how they created a chemically enriched universe. And that was the crucial Goldilocks condition for building complex molecules that could form new types of astronomical objects, such as our Earth and its moon.

CHAPTER 3

Molecules and Moons: Threshold 4

In truth there are only atoms and the void.
— DEMOCRITUS

You're on Earth. There's no cure for that.
— SAMUEL BECKETT, *ENDGAME*

From Stardust to Molecules

So far we have seen how violent processes using extreme energies and guided by the universe's basic operating rules created galaxies, stars, and new elements. They did so with the cosmological equivalent of chain-saw sculpture, and gravity is a virtuoso chain-saw sculptor. Near stars, this rough sculpture provided new environments in which more delicate sculpture became possible. To understand these new types of structures, we need to move from very large things to very small things. We need to focus on the relationships between atoms.

Chemical complexity depends on tiny flows of electromagnetic energy that can do the nano-work of rearranging single atoms and molecules. But such delicate flows of free energy are common only in sheltered, and rare, Goldilocks environments. High temperatures will blast molecules and atoms apart, so chemical complexity is impossible within stars. But chemical complexity does require some energy, so it is also impossible in the dead zone of deep space. The ideal environment seems to be near a star, but not too near, in regions with sustained but gentle flows of free energy.

We humans feel gravity, but in the nano-world where atoms hang out, gravity is not so important. It doesn't even matter much to small things like bacteria or water striders, which are much more concerned with local electric charges or the surface tension of water, respectively. At the scale of molecules, the electromagnetic force rules. This is the force that glues atoms and molecules together and pries them apart. Molecules and atoms move through a sticky world of electromagnetic hooks, probes, lures, and lassos.

Chemistry began inside galactic dust clouds as they filled with new elements. Even today, about 98 percent of the mass of interstellar dust clouds consists of hydrogen and helium. But sprinkled among the hydrogen and helium atoms are atoms of all the other elements in the periodic table. Confusingly, astronomers term all elements heavier than helium *metals*. So they tell us that as more and more large stars died, the universe got more metallic. Similarly, we could say that our sun is more metallic than earlier generations of stars because it contains more metals.

Spectroscopes can tell us what elements are present in galactic clouds and in what amounts. Spectroscopes can also identify molecules, clusters of atoms bound together by electromagnetic forces. They can tell, for example, if the cloud contains molecules of water or ice or molecules of silicates, which consist primarily of silicon and oxygen and make up most dust and rocks on Earth. We now know that there are many simple molecules in galactic dust clouds, and they include some, such as amino acids (the building blocks of proteins), that are crucial for life on Earth.

Chemistry is the discipline that explores how electromagnetic forces build molecules and how atoms combine and recombine to form the kaleidoscopic material diversity of our world.

Chemical Trysts: How Atoms Combine

Atoms are tiny. You could pack a million carbon atoms into the dot at the end of this sentence. But don't think of atoms as solid balls of matter. They consist almost entirely of empty space. Each has a tiny nucleus in its center made up of protons (with positive charges) and neutrons (which have no charge) bound together by the strong nuclear force. The rest of the atom is mostly empty. Orbiting the nuclei at huge distances are clouds of electrons, roughly one to each proton in the nucleus. Early in the twentieth century, Ernest Rutherford, one of the pioneers of modern nuclear physics, described the nucleus of an atom as "the fly in the cathedral."

The scale Rutherford suggests is about right. But he was writing before the evolution of modern quantum physics, which showed that his metaphor is also misleading. Electrons are minuscule, with about $1/1836$ the mass of a proton. Quantum physics showed that we can never pin down their exact speed or position. We can tell where an electron *probably* is, but never *exactly* where it is, because any attempt to locate it will require the use of energy (imagine shining a flashlight on it), and electrons are so light that the energy used to detect them will alter their speed and trajectory. This is why quantum physicists map orbiting electrons onto a sort of "probability mist" that thickens at certain distances from the nucleus and thins out at others. The probability mist pervades most of the atomic cathedral and can seep beyond its outer walls.[1]

Chemistry is all about the trysts and the wars inside these probability mists. And there's a lot going on. Bonds are formed and broken between protons and electrons, old ties are ended, new relationships are started, and the result is the emergence of entirely new forms of matter. Driving all this activity is the simple fact that electrons have negative charges that repel each other but attract them to the positive charges of protons, either in their home atom or in neighboring atoms. Chemists study these flirtations and rivalries and the liaisons and tensions they create as electrons hitch up to neighboring atoms to form molecules linking several atoms, sometimes millions or even billions of them, into structures more complex than the most complex star. Each molecular pattern has distinctive emergent properties, so the possibilities of chemistry seem endless. Nevertheless, the courtships have their own operating rules (sometimes as perverse as the rules of a human courtship), and these govern how the electromagnetic force can build chemical complexity.

Electrons are the key players. Like human lovers, electrons are unpredictable, fickle, and always open to better offers. They buzz around protons in distinct orbits, each associated with a different energy level. Wherever possible, electrons head for the orbits closest to an atom's nucleus, which require the least energy. But the number of spaces in each orbit is limited, and if no places are left in the inner orbits, they have to settle for places in an outer orbit. If that orbit has just the right number of electrons, everyone is happy. This is the situation of the so-called noble gases, such as helium or argon, which you find over on the right-hand side of the periodic table. They don't combine with other atoms because they are more or less content with the status quo.

But if the outer orbits of an atom are not filled, that creates awkwardness, problems, and tensions, and the endless jostling for position this causes can explain a lot of chemistry. Some electrons jump ship and head for neighboring atoms. If they do that, the atom they left will have lost a negative charge, so it may pair up with an atom that has an extra electron to form an ionic bond. This is how salt is formed from atoms of sodium, whose outermost electron is usually willing to jump, and chlorine, which is often looking for an extra electron to fill up its outer orbit. Sometimes, electrons will feel most comfortable when they are orbiting two nuclei, so the nuclei effectively share their charges in a covalent bond. This is how atoms of hydrogen and oxygen combine to form water molecules. But the molecule they form is lopsided, with two smallish hydrogen atoms glomming on to one side of a larger oxygen atom. That odd shape distributes negative and positive charges unevenly over the molecule's surface and confuses hydrogen atoms, which often get attracted to the oxygen atoms in neighboring molecules. That attraction explains why water molecules can stick together in droplets, exploiting these weak hydrogen bonds. Hydrogen bonds play a fundamental role in the chemistry of life because they account for much of the behavior of genetic molecules such as DNA. In metals, electrons behave very differently. Vast crowds of electrons cruise among metallic nuclei, and that explains why metals are so good at conducting electric currents, which are really huge streams of electrons.

Carbon, with six protons in its nucleus, is the Don Juan of these atomic romances. It normally has four electrons in its outer orbit, but there is room enough there for eight, so you can make a carbon atom happy by removing four electrons from its outer shell, by adding four electrons, or by letting it share four electrons with another atom. This gives it a lot of options, and that is why carbon can form complicated molecules with rings, chains, and other exotic shapes. Its virtuosity explains why carbon is so important to the chemistry of life.

The basic rules of chemistry seem to be universal. We know this because spectroscopes show that many of the simple molecules we find on Earth also exist in interstellar dust clouds. But interstellar chemistry seems to be fairly simple; no interstellar molecules detected so far have more than about a hundred atoms. And that is no surprise. After all, in space, atoms are far apart, so it's difficult for them to hitch up with one another. Besides, temperatures are chilly, so there is little of the activation energy that is needed to nudge atoms into long-term partnerships. What is most striking about interstellar chemistry is that it can generate not only the simple molecules from which planets are formed, such as water and silicates, but also many of the basic molecules of life, such as amino acids, the components of proteins. Indeed, we now know that simple organic molecules are common in the universe, and that increases the likelihood that life exists beyond planet Earth.

Threshold 4: From Molecules to Moons, Planets, and Solar Systems

Simple chemical molecules orbiting young stars created the Goldilocks conditions for our next threshold of increasing complexity, because they provided the building blocks for entirely new astronomical objects: planets, moons, and asteroids. Planetary bodies were chemically richer than stars, and much cooler, so they offered ideal Goldilocks environments for complex chemistry. And on at least one planet (our own), and probably on many more, that chemistry would eventually generate life.

For a long time, people knew about just one solar system. But in 1995, astronomers identified *exoplanets,* planets orbiting other stars in our galaxy. They did this by detecting tiny wobbles in the motions of stars or tiny variations in their brightness as planets crossed in front of them. Since then, we have learned that most stars have planets, so there could be tens of billions of planetary systems of many different types just in our galaxy. By the middle of 2016, astronomers had identified more than three thousand exoplanets. In the next decade or two, the study of other planetary systems should give us a better sense of the most common configurations. Soon, we should be able to study their atmospheres, which may give us a sense of how many could be life-friendly. We already know that many are about the same size as Earth, and many orbit at the right distance from their stars to have liquid water—a crucial ingredient for life.

The discovery of exoplanets tells us that, like threshold 3, threshold 4 has been crossed many times, and it may have been crossed for the first time quite early in the universe's history around a star we will probably never detect. But we now know quite a lot about what the crossing of this threshold looks like.

The formation of planetary systems is a messy and chaotic process, a by-product of star formation in chemically enriched regions of space. Billions of years after the big bang, interstellar space was filled with clouds of matter containing many different chemical elements. Hydrogen and helium still made up almost 98 percent of those clouds, but it was the remaining 2 percent that made the difference. As in the early universe, gravity liked to make these clouds more clumpy. In our region, gravity may have been helped by a nearby supernova explosion that shook things up and started the contraction of a huge cloud of gas and dust about 4.567 billion years ago. The supernova left its calling card in distinctive radioactive materials that show up in meteorites within our solar system.

As it contracted, the dust cloud broke up into multiple solar nebulae, one of which formed our sun. The sun gobbled up 99.9 percent of all the matter in its dust cloud. But what interests us now is the leftovers, the rings of debris orbiting the young sun. As gravity shrank the solar nebula, its swirling mass of gas, dust, and ice particles spun faster and faster until centrifugal forces flattened it like pizza dough to create the thin plane of today's solar system. We can now observe such protoplanetary disks in nearby regions of star formation, so we know they are very common.

Two processes turned a spinning disk of matter into planets, moons, and asteroids. The first was a type of chemical sorting. Violent bursts of charged particles from the young sun, known as the solar wind, blasted lighter elements, such as hydrogen and helium, away from the inner orbits to create two distinct regions. The outer regions of the young solar system, like most of the universe, consisted mainly of the primordial elements, hydrogen and helium. But the inner regions, where the rocky planets—Mercury, Venus, Earth, and Mars—would form, lost so much hydrogen and helium that they had a rare chemical diversity. Oxygen, silicon, aluminum, and iron make up over 80 percent of Earth's crust, with elements such as calcium, carbon, and phosphorus playing lesser roles. On Earth, hydrogen plays only a medium-size role, and helium is hardly ever sighted.

The second process that formed our solar system was accretion. Within different orbits around the young sun, bits of matter slowly gathered together. In the gassier outlying regions, this was probably a fairly gentle process. Gravity collected matter into huge gassy planets, such as Jupiter and Saturn, that consisted mostly of hydrogen and helium with a thin sprinkling of dust and ice. In the inner regions, though, accretion was a more violent and chaotic process, because here, a lot more matter was solid. Particles of dust and ice stuck together to form small blobs of rock and ice, which careened around, sometimes smashing each other into pieces, sometimes sticking together to form larger objects. Eventually, even larger objects appeared, like meteors and asteroids, and within each orbit, these smashed into each other or merged to form objects so large that their gravity could sweep up most of the remaining debris. Eventually, these processes generated the planets we see today, spaced out in distinct orbits around the sun.

Such an account gives little sense of the chaos and violence of accretion. Some objects crossed orbits, knocking young planets and moons out of alignment or smashing them to pieces. The vast protoplanet of Jupiter may have migrated inward, its gravitational pull breaking up any would-be planet that was forming in what is now the asteroid belt. Uranus's odd tilt and rotation are probably the result of a violent collision with another large body. And the jagged forms of many asteroids are the scars of brutal collisions early in the history of our solar system.

Collisions continued for a long time, even as the solar system stabilized. Indeed, our own moon was probably formed from a collision between the young Earth and a Mars-size proto-planet (Theia) about one hundred million years after the birth of the solar system. That collision sent huge clouds of matter into orbit around Earth, where they probably circled like the rings of Saturn (which might also be the debris from a smashed-up moon) until they accreted to form our moon.

Within fifty million years, our solar system had acquired the basic shape it has today, and since then it has proved quite stable. The billions of planetary systems in our universe probably formed in similar ways, though they exist in a great variety of different configurations. But all planetary bodies are cooler than stars, and chemically richer and more diverse, and that's why they provided Goldilocks conditions that allowed the building of new forms of complexity. Eventually, at least one of these objects, and probably many more, generated life.

Planet Earth

Our solar system lies in the galaxy we call the Milky Way in a stellar suburb on one of the Milky Way's spiral arms, the Orion spur. The Milky Way is one member of a group of about fifty galaxies, known, unromantically, as the Local Group. The Local Group lies in the outer regions of the Virgo Cluster, which has about a thousand galaxies. This is part of the Local Supercluster, which includes hundreds of groups of galaxies. It would take you one hundred million years traveling at the speed of light to cross it. In 2014, it was found that the Local Supercluster is part of a vast cosmic empire with perhaps a hundred thousand galaxies, and to cross that would take you four hundred million years traveling at the speed of light. This empire is the Laniakea (Hawaiian for "immeasurable heaven") Supercluster. At present, this is the largest structured entity we know of in the universe. We assume that Laniakea is built around a scaffolding of dark matter whose gravitational pull holds all these galaxies together as the universe expands.

Now we must travel back to the suburbs of Laniakea, to our own local group, our own galaxy, and out to the Orion spur, where we find our own sun and planet Earth. After Earth formed by accretion, one final display of chain-saw sculpture gave it its distinctive inner structure. Geologists call this process *differentiation*.

The young Earth heated up and melted. It was heated by the violent collisions of accretion, by the presence of radioactive elements (created in the supernova that provided much of the material for our solar system), and by increasing pressure as it grew in size. Eventually, the young Earth was so hot that much of it melted into a gooey sludge, and as it liquefied, its different layers sorted themselves by density, giving it the structure it has today.

The heavier elements, mainly iron and nickel and some silicon, sank through the hot sludge to the center to form Earth's metallic core. As Earth spun, the core generated a magnetic field that shielded the surface from the damaging charged particles of the solar wind. Lighter rocks, such as basalts, gathered above the core to form a second layer, a three-thousand-kilometer-deep region of semimolten rock mixed with gas and water that's known as the mantle. This is where the lava belched up by volcanoes comes from. The lightest rocks, many of them granites, floated to the surface, where they cooled and solidified to form a third layer: the eggshell-thin stratum known as the crust, which is covered today by oceans and continents. Beneath the oceans, the crust is sometimes just five kilometers thick, but under the continents, it can be up to fifty kilometers thick. The crust is particularly interesting chemically. In it, you can find solids, liquids, and gases, and it was repeatedly heated and cooled by volcanoes, asteroid impacts, the harsh glare of the young sun, and the eventual condensation of Earth's first oceans. Here and in the mantle, heat and the circulation of elements generated perhaps two hundred and fifty new minerals.[2] Gases, including carbon dioxide and water vapor, bubbled up from the mantle through volcanoes and cracks in the surface to form a fourth layer: Earth's first atmosphere. The crust and atmosphere were also enriched by gases, water, complex molecules, and other materials brought in by asteroids and comets.

The hot, molten core kept the young Earth dynamic, as energy from the center worked its way through the planet, heating and churning up its upper layers to create circulating currents of soft rock in the mantle and a surface dotted with volcanoes. Heat from the core still drives change in the upper levels of planet Earth. Today, we can track the movement of the surface using GPS systems, and we know that crustal plates on the surface move around at about the speed that your fingernails grow; the fastest of them cruise at about twenty-five centimeters a year.

Geologists divide Earth's history into subdivisions, the largest of which is the eon. The first is the Hadean ("hell-like") eon. This lasted from when Earth formed to about four billion years ago, when the Archean eon began. If you'd visited during the Hadean eon, you'd have found a planet still affected by the demolition derby of accretion. Gouges and tears on the surface of the moon and other planets show that between 4.0 and 3.8 billion years ago, the inner solar system was subjected to a massive pummeling by asteroids and other stray objects. This is known as the Late Heavy Bombardment, and it was probably caused by shifts in the orbits of Jupiter and Saturn, which sprayed objects at random around the young solar system. Today, most of the asteroids live between Jupiter and Mars, so they may be the bricks and struts of a planet that was never built because of Jupiter's disruptive gravitational tug. At present, we know of some three hundred thousand asteroids. Though most are small, that's a lot of stray matter with which to bombard the inner planets.[3]

Studying Earth: Seismographs and Radiometric Dating

Despite what Hollywood might have us believe, we cannot dig deep into Earth. The deepest dig so far is about twelve kilometers, which is about 0.2 percent of the distance to Earth's core. That hole was drilled in the Kola Peninsula in the far northwest of Russia as part of a geological investigation. We know about the interior because of another neat scientific trick, the geologist's equivalent of an X-ray. Earthquakes generate tremors that travel through Earth's interior. Seismographs measure those tremors at different places on the surface. By comparing results from different regions, you can figure out how fast and how far tremors have traveled through the interior. We also know that different types of tremors travel at different speeds through different materials, and some travel only through solids, while others can travel through liquids as well. So tracking these tremors with multiple seismographs can tell us a lot about Earth's interior.

Determining Earth's age and the many other dates scattered throughout the modern origin story became possible only in the second half of the twentieth century, and it depended on some very clever science.

The first steps toward a modern history of planet Earth were taken in the seventeenth century. That's when some of the pioneers of modern geology realized that it might be possible to determine the *order* of events in Earth's history, even if no one had any idea of exactly *when* things happened. In the seventeenth century, a Danish priest who lived in Italy, Nicholas Steno, showed that by carefully studying sedimentary rocks, you could determine the order in which different rock strata had been laid down. All sedimentary rocks are built up layer by layer, so we know that the oldest layers are normally the lowest ones. Anything cutting through them had to have been younger.

Early in the nineteenth century, an English surveyor, William Smith, showed that identical suites of fossils appeared in rock formations in different places. On the reasonable assumption that similar fossils must have come from about the same time, you could identify strata around the world that had been laid down at the same time. Taken together, these principles allowed nineteenth-century geologists to create a relative timeline for Earth's history. That timeline still lies behind modern geological dating systems, and it begins with the Cambrian period, the first period whose strata contained fossils visible to the naked eye.

But no one knew *when* the Cambrian period had occurred, and many geologists despaired of ever finding absolute dates for different strata. In 1788, James Hutton wrote: "We find no vestige of a beginning, no prospect of an end."[4] Even early in the twentieth century, the only way to give an absolute date to an event was to find a written record that mentioned it. And that meant, as H. G. Wells pointed out when he tried to write a modern origin story just after World War I, that absolute timelines could reach back no farther than a few thousand years.

Though H. G. Wells didn't know it, some of the discoveries that would eventually provide better dates had already been made. The key was radioactivity, a form of energy discovered by Henri Becquerel in 1896. In atoms with large nuclei, such as uranium, the repulsive power of lots of positively charged protons can destabilize the nucleus until, eventually, it breaks down spontaneously, emitting high-energy electrons or photons or even whole helium nuclei. As chunks of the nucleus are ejected, the element is transformed into different elements with fewer protons. For example, uranium eventually breaks down to lead. In the first decade of the twentieth century, Ernest Rutherford realized that, even if you could not tell when a particular nucleus was about to break apart, radioactive breakdown was a very regular process when averaged over billions of particles. Every isotope of the same element (isotopes have the same number of protons but different numbers of neutrons) breaks down at different but regular rates, so it is possible to determine precisely how long it will take for half of the atoms in a given isotope to decay. For example, the half-life of uranium-238 (with 92 protons and 146 neutrons) is 4.5 billion years, while uranium-235 (with 92 protons and 143 neutrons) has a half-life of 700 million years.

Rutherford realized that radioactive breakdown could provide a sort of geological clock if you could measure how much a given sample had decayed. In 1904, he tried to measure the breakdown of a sample of uranium and came up with a figure of about five hundred million years for the age of Earth. The basic idea was right, but his estimate of Earth's age was controversial because it was much older than the accepted age of less than one hundred million years.

Over time, an increasing number of geologists began to agree that Earth might be much older than they had once thought. But the technical problems of measuring radioactive breakdown were formidable. They were solved only in the late 1940s, using methods developed as part of the Manhattan Project, which had manufactured the first atomic bomb. To make the bomb, it was necessary to separate different isotopes of uranium in order to produce purified samples of uranium-235. An American physicist, Willard Libby, helped develop the techniques for separating and measuring different isotopes of uranium, and these would prove crucial in the task of measuring radioactive breakdown.

In 1948, Libby's team managed to give accurate dates for material from the tomb of the pharaoh Zoser, which had been provided by the Metropolitan Museum.[5] They used carbon-14, a radioactive isotope of carbon that has a half-life of 5,730 years, which makes it extremely useful when studying organic materials such as wood. Different radioactive materials worked at different scales and with different materials. For geologists, the decay of uranium to lead was particularly valuable, and the fact that different isotopes of uranium decay at different rates allowed cross-checking.[6] In 1953, Clair Patterson dated the age of an iron meteorite using the decay of uranium to lead. He made the correct assumption that meteorites were made up of primordial material from the young solar system and could therefore provide an age for the entire solar system. His measurements suggested Earth was about 4.5 billion years old, much older than Rutherford had estimated. Patterson's date still stands today.

Along with radiometric dating techniques, there have emerged other dating techniques that can be used to check each other. Dates within recent millennia can sometimes be determined by counting the annular rings of ancient trees such as bristlecone pines, which can live for several thousand years. Astronomers use their own techniques for dating the history of the universe, and biologists have found that DNA evolves at a reasonably regular pace, so you can roughly date when two species diverged from a common ancestor by measuring differences in their genomes. Such techniques, based on careful study of processes such as radioactive decay, as well as the development of new instruments for measuring them precisely, have given us the timelines around which the modern origin story is built.

So far, we have watched complexity increasing in entities that are interesting but not alive. Now we reach one of the most fundamental of all our thresholds: the appearance of life. With life, we encounter an entirely new type and level of complexity and a whole series of new concepts, including information, purpose, and even, eventually, consciousness.

PART II

Biosphere

CHAPTER 4

Life: Threshold 5

Life and Information: A New Type of Complexity

> *I spent the afternoon musing on Life. If you come to think of it, what a queer thing Life is! So unlike anything else, don't you know, if you see what I mean.*
> — P. G. WODEHOUSE, *MY MAN JEEVES*

> *What lies at the heart of every living thing is not a fire, not warm breath, not a "spark of life." It is information, words, instructions.... If you want to understand life, don't think about vibrant, throbbing gels and oozes, think about information technology.*
> — RICHARD DAWKINS, *THE BLIND WATCHMAKER*

Life as we know it arose from exotic chemistry in the element-rich environments of the young planet Earth almost four billion years ago. If life exists elsewhere, it might look so strange that we wouldn't recognize it. But on planet Earth, life is built from billions of intricate molecular nanomachines. They work together inside protective bubblelike structures we consider the building blocks of life—the basic structural, functional, and biological units of all known living organisms. These protected bubbles are called cells, from the Latin *cella*, meaning "small room."

Cells are the smallest units of life that can replicate independently. They survive by tapping delicate flows of nutrients and free energy from their surroundings.

Life has had a colossal impact on our planet because living organisms make copies of themselves that can multiply, spread, proliferate, and diversify. Over four billion years, a colossal army of living organisms has transformed Earth and created the biosphere: a thin layer at the planet's surface made up of living organisms and everything shaped or altered or left behind by living organisms.

The spooky thing about life is that, though the inside of each cell looks like pandemonium—a sort of mud-wrestling contest involving a million molecules—whole cells give the impression of acting with purpose. Something inside each cell seems to drive it, as if it were working its way through a to-do list. The to-do list is simple: (1) stay alive despite entropy and unpredictable surroundings; and (2) make copies of myself that can do the same thing. And so on from cell to cell, and generation to generation. Here, in the seeking out of some outcomes and the avoidance of others, are the origins of desire, caring, purpose, ethics, even love. Perhaps even the beginnings of meaning, if that means the ability to discriminate between the significance of different events and signs. What is the meaning of this great white shark cruising behind me?

The appearance (or, perhaps, illusion) of purposefulness is new. It is *not* a feature of the other complex entities we have seen so far. Would it mean anything to say that stars have a purpose? Or planets, or rocks? Or even the universe? Not really, at least not within the conventions of the modern origin story. But living things are different. They don't accept entropy's rules passively; instead, like stubborn children, they push back and try to negotiate. They don't just lock structures in place, like protons or electrons. They don't live off stores of energy, like stars, which munch their way through a larder of protons that was well stocked at their birth and then fall apart when the larder is empty. Living organisms constantly seek out new flows of energy from their environments in order to maintain themselves in a state that is complex but unstable. This is not the behavior of rocks; it is that of a bird on the wing. Living organisms stay airborne (thermodynamically speaking) by taking in free energy to drive the elaborate chemistry that rearranges atoms and molecules in the patterns needed to keep them alive. When they can no longer pay entropy's energy taxes, they crash.

Energy and life! In Australia, I remember watching my own children transform the energy in Vegemite sandwiches into the violent energy of motion as they roared around the garden. We can even measure the rate at which free energy (perhaps from a Vegemite sandwich) flows as it is transformed into talking energy, running energy, and, eventually, heat energy, with entropy increasing at each step. The average human takes in about 2,500 calories each day, about 10.5 million joules (a measure of work or energy; a calorie represents about 4,184 joules). Divide this by the 86,400 seconds in a day, and an individual mobilizes about 120 joules every second. This is the "power rating" of a human being: 120 watts, just slightly greater than the power rating of many traditional lightbulbs.[1]

Life, with its never-ending attempts to push back against entropy, represents a new type and level of complexity. Complexity theorists sometimes describe entities at this level as complex adaptive systems. Unlike the complex physical systems we have seen so far, the components of which behave in ways that can usually be predicted from the universe's basic operating rules, the components in complex adaptive systems seem to have a will of their own. They appear to follow additional rules that are harder to detect. Indeed, complex adaptive systems, such as bacteria, your dog, or multinational companies, act as if every component is an agent with a will of its own, so each component is constantly adjusting to the behavior of many other components. And that yields extremely complex and unpredictable behaviors.[2]

In using the word *agent*, I have smuggled in a new idea that will become increasingly important: the idea of information. If agents react to other agents, they are reacting to information about what is happening around them, including information about what other agents are doing. If we imagine information as a character in our modern origin story, we should think of it as working undercover or in disguise, manipulating events but staying out of the spotlight. Energy *causes* change, so you can usually see it at work, but information *directs* change, often from the shadows. As Seth Lloyd puts it: "To do anything requires energy. To specify what is done requires information."[3]

In its most general form, information consists of rules that affect outcomes by limiting possibilities. One of the most famous definitions of *information* is "a difference which makes a difference."[4] Rules determine which changes out of all conceivable options are actually possible at a given time and place, and that makes a difference. Information begins with the laws of physics, the basic operating system of our universe. The laws of physics steer change down particular pathways, like the pathways by which gravity created the first stars. Information in this very general sense limits what is possible, so it reduces randomness. This is why more information seems to mean less entropy, less potential for the disorder that entropy loves. This is universal information: the rules built into every smidgen of matter and energy. No one needed to tell gravity what to do; it just got on with the job.

In colloquial usage, though, the term *information* means more than rules. It means rules that are *read* by some person or agent or thing—in fact, by some complex adaptive system. This sort of information arises because many important rules are *not* universal. Like the laws of human societies, they change from place to place and moment to moment. As the universe evolved, new environments appeared, such as deep space, galactic dust clouds, and the surfaces of rocky planets. These environments had their own local rules that were *not* universal. Local rules have to be read or decoded or studied, just as you might want to learn which side of the road locals drive on before visiting Mongolia (the right, by the way).

Complex adaptive systems can survive only in very specific environments, so they need to be able to read or decode local information as well as the universal rules. And that's new. All forms of life require mechanisms to interpret local information (such as the presence of different chemicals or local temperatures and acidity levels) so they can respond appropriately (*Should I hug it or eat it or run?*). The philosopher Daniel Dennett writes: "Animals are not just herbivores or carnivores. They are... *informavores*."[5] In fact, all living organisms are informavores. They all consume information, and the mechanisms they use for reading and responding to local information—whether they are eyes and tentacles or muscles and brains—account for much of the complexity of living organisms.

Local environments are unstable, so living organisms must constantly monitor their internal and external environments to detect significant changes. And as organisms increase in complexity, they need more and more information, because more complex structures have more moving parts and more links between their parts. The bacterium *E. coli*, which is probably flourishing in your intestines as you read this, allocates about 5 percent of its molecular resources to movement and perception, but in your body, *most* organs are devoted, directly or indirectly, to perception or motion, from brains to eyes to nerve tissues and muscles.[6] Modern science is at the extreme end of a vast spectrum of information-gathering-and-analyzing systems that begin with the simple sensors of the earliest single-celled organisms.

Entropy, of course, keeps a beady eye on all of this. If more complexity means more information, then when you increase complexity and information, you are *reducing* entropy and its accompanying uncertainty or disorder. And entropy will notice. Entropy is rubbing its hands at the thought of the energy taxes and fees it can levy as complexity and information increase.[7] Indeed, some have argued that entropy actually likes the idea of life (and may encourage it to appear in many parts of the universe), because life degrades free energy so much more efficiently than nonlife.

Explaining the origins of life on Earth and trying to figure out if something similar might have emerged elsewhere in our universe are among the most difficult problems facing modern science. At the moment, we know of only one planet with life. Astrobiologists are searching for life elsewhere through the Search for Extraterrestrial Intelligence (SETI) program, which began in 1960, but so far they have found none. For now, we are confined to studying the origins of life on Earth. Even that is extraordinarily difficult, as it means trying to determine what was happening on our planet almost four billion years ago, when Earth was very different.

Defining Life

Having only one sample makes it difficult even to know what life *is*. What distinguishes life from nonlife? Life is as hard to define as complexity or information, and there seems to be a murky border zone between life and nonlife.

Most modern definitions of life on Earth would include the following five features:

1. Living organisms consist of cells enclosed by semipermeable membranes.

2. They have a metabolism, mechanisms that tap and use flows of free energy from their surroundings so they can rearrange atoms and molecules into the complex and dynamic structures they need to survive.
3. They can adjust to changing environments by homeostasis, using information about their internal and external environments and mechanisms that allow them to react.
4. They can reproduce by using genetic information to make almost exact copies of themselves.
5. But the copies differ in minute ways from the parents, so, over many generations, the features of living organisms slowly change as they evolve and adapt to changing environments.

Let's take each of these features in turn.

All living things on Earth consist of cells. Each cell contains millions of complex molecules that react with one another many times every second as they push their way through a watery, salty chemical sludge full of proteins in the gooey realm known as the cytoplasm. The cytoplasm is bounded by a sort of chemical fence, the cell membrane, that controls what comes in and goes out. Like the walls of a medieval city, the membrane has gates and guards that decide which molecular travelers can enter and when. Cells really are like cities. In a book on cells, Peter Hoffmann writes:

There is a library (the nucleus, which contains the genetic material), power plants (mitochondria), highways (microtubules and actin filaments), trucks (kinesin and dynein), garbage disposals (lysosomes), city walls (membranes), post offices (Golgi apparatus), and many other structures fulfilling vital functions. All of these functions are performed by molecular machines.[8]

All living organisms depend on carefully managed flows of free energy. Stop the flow, and they die, like a besieged city starved into submission. But if the flow is too violent, they will also die, like a city under aerial bombardment. So flows of energy need to be managed with great delicacy. Usually, cells take in and use energy in tiny doses, electron by electron or proton by proton. Though small enough not to be disruptive, these flows are large enough to provide the activation energies needed to drive lots of interesting chemistry. Etymologically, the word *metabolism* comes from the word meaning "change." It's a reminder that cells never stand still. Like birds in flight, they use flows of energy to keep adjusting to ever-changing environments.

Living organisms must constantly monitor and adjust to changes in their environments. This constant adjustment is known as preserving homeostasis. To maintain some sort of equilibrium in changing surroundings, cells must continually access, download, and decode information about their internal and external environments, decide on the best response, and then respond. The word *homeostasis* means "standing still," which is the opposite of "change." But it makes sense if you think of standing still in the never-ending molecular hurricane of the cell's environment.

Impressive as these abilities are, they would be of little interest if living organisms appeared and vanished like spray on an ocean wave. And that may be what has happened on some planets around some stars, and perhaps even early in Earth's history. But today on planet Earth, living organisms don't just stand up in the hurricane of change and entropy. They also make copies of themselves, so that when particular cells fall down (and eventually they will all fall down), others can take their place. Reproduction is the ability to make viable copies of cells. Reproduction means that the *template* for making an organism (its *genome*, in modern terminology) can survive even after individuals have died. Like an instruction manual, the genome stores information about the proteins needed to build a copy of the parent as well as some basic assembly rules. Today, most of this information is stored in molecules of DNA. But early in the history of life on Earth, it was probably stored in RNA, a molecular cousin to DNA that still does a lot of heavy lifting inside cells.

Though the templates are more or less immortal, the copying process is not perfect. This is good news, because it means the templates can slowly change as a result of tiny copying errors, and that is the key to adaptation and evolution. Tiny genetic changes give life its remarkable resilience because they allow species to adapt to their environments by randomly creating slightly different templates. As environments change, so, too, do the rules that determine which templates will survive and which will perish.

This is the mechanism Charles Darwin described as *natural selection*. Natural selection is a fundamental idea in modern biology because it is an extraordinarily powerful driver of increasing complexity. Natural selection filters out some genetic possibilities, allowing only those compatible with local rules. So natural selection is a ratchet, like the fundamental laws of physics, because it locks nonrandom patterns in place. But in the biological realm, it is the local rules of particular environments, not the universal rules of physics, that determine what survives. And the biological rules are much more persnickety. Don't expect a giraffe to survive underwater.

Like the mechanisms that generated the universe's first structures, natural selection links necessity and chance. Variation provides multiple possibilities; natural selection uses local rules to pick out those that will work under local conditions. Here is how Darwin put it in *The Origin of Species:*

> Can it . . . be thought improbable [that] variations useful
> in some way to each being in the great and complex
> battle of life, should sometimes occur in the course of

thousands of generations? If such do occur, can we doubt (remembering that many more individuals are born than can possibly survive) that individuals having any advantage, however slight, over others, would have the best chance of surviving and of procreating their kind? On the other hand, we may feel sure that any variation in the least degree injurious would be rigidly destroyed. This preservation of favourable variations and the rejection of injurious variations, I call Natural Selection.[9]

Darwin's idea, when linked to a modern understanding of genetics and heredity, explains life's creativity, its ability over many generations to explore possibilities, tap new energy flows, and construct new types of structures. It explains how, in the biological realm, structures of staggering complexity can emerge through repetitive algorithmic processes as they are filtered out from myriad variations, step by step and generation by generation, over millions and billions of years.

The idea of natural selection shocked Darwin's contemporaries, because it seemed to do away with the need for a creator god.[10] And that idea was fundamental to the Christian origin story that most people accepted in Victorian England. Even Darwin was worried, and his wife, Emma, feared she and Charles would end up in different places in the afterlife. But the mechanism Darwin described really does seem to be fundamental to the history of life. Let finches breed on one of the Galápagos Islands that Darwin visited in his youth. If this island's trees produce nuts with tough shells, over time those finches with beaks that can crack the shells most efficiently will survive better and produce more offspring than others. Wait a few generations, and you will find all the finches on this island have this type of beak. Over time, as some individuals are selected by "nature" (in fact, by the rules of the local environment), a new species will eventually emerge. Here, as Darwin showed, is the basic mechanism of biological evolution. This is Darwin's complexity ratchet; this is how life builds more and more complex things, step by step.

The Goldilocks Conditions for Life

How did life first sputter into motion somewhere in the rich and varied Goldilocks environments of the young Earth?[11]

What Darwin did not know was that mechanisms similar to natural selection, in which random changes are filtered out by local rules, can also work in rough-and-ready ways in a world without life. Where there are complex mixtures of chemicals and plenty of free energy, molecules can arise that encourage the formation of other molecules and eventually create the molecules the reaction started out with. This is an autocatalytic cycle, a reaction whose components enable, or *catalyze,* the production of other components of the cycle, including its original ingredients, so the cycle can repeat itself. Fire up one of these cycles, and it will produce its components in larger and larger quantities as it extracts more and more food energy until it starts starving other, less successful reactions. The cycle may even modify itself slightly if new types of food appear. This is beginning to look like the survival of the most successful chemical reaction. So here we already have something a bit lifelike, something that can persist and reproduce by tapping energy from its surroundings. "Before we can have competent reproducers," writes Daniel Dennett, "we have to have competent persisters, structures with enough stability to hang around long enough to pick up revisions."[12] This idea of chemical evolution will help us explain, at least in general terms, how the preconditions for life emerged on the young Earth.

Chemical evolution can take place only in an environment

that allows rich chemical experimentation. And such environments are extraordinarily rare. So what are the Goldilocks conditions for chemical experimentation? And why did the young Earth exhibit so many of them?

First, our solar system is in the right part of the Milky Way galaxy. Stars in the galaxy's outer suburbs have thin, chemically impoverished clouds of chemicals to work with. Stars too close to the galaxy's central business zone are battered by shock waves from the violent outbursts of black holes that lie at its core. Our solar system is in just the right place. Its orbit is about two-thirds of the way from the center of the Milky Way, in the middle of our galaxy's "habitable zone."

Second, chemistry works well only at lower temperatures. The early universe was too hot for atoms to combine into molecules. So is the interior of stars. Rich chemistry is possible within only a narrow range of fairly low temperatures, and you find these in habitable zones that are close to stars but not too close. Our Earth's orbit is in about the middle of our sun's habitable zone. Venus and Mars orbit at the inner and outer edges, respectively, of our system's habitable zone. But we are learning that some moons farther away from the sun, such as Saturn's moon Enceladus, may also have internal furnaces and chemistries that make them life-friendly. In 2017, scientists found that the oceans of Enceladus produce hydrogen, that gas that provided food for some of the earliest organisms on planet Earth.[13]

A third Goldilocks condition for rich chemistry is the presence of liquids. In gases, atoms zoom about like hyperactive kids, so it's hard to keep them still enough to hitch up with other atoms. In solids, you have the opposite problem: atoms are locked in place. But liquids are like ballrooms, and liquid water, with its whispering hydrogen bonds, offers the best ballroom of all. Atoms can cruise, waltz, and tango, and it's not too hard for electrons to change partners if they spot something more attractive. The presence of liquids depends on chemistry, temperature, and pressure. There is a narrow range of temperatures in which water exists in liquid form (most water in the universe is in the form of ice). But at these same temperatures, you can also find gases and solids, which makes for very interesting chemical possibilities. So, we should expect the most interesting chemistry to be on planets whose average surface temperatures lie roughly between zero and one hundred degrees Celsius, the freezing and boiling points, respectively, for water. That's rare, but our Earth happens to be at just the right distance from the sun to have liquid water.

A fourth Goldilocks condition for rich chemistry is chemical diversity. It's no good having the right temperature if you've got only hydrogen and helium to work with. And today, even in the chemically rich regions within galaxies, hydrogen and helium still make up 98 percent of all atomic matter. What chemistry needs is those rare environments in which the other elements of the periodic table are more common. In our solar system, such diversity can be found only on the rocky planets close to the sun, because the young sun boiled away much of the hydrogen and helium from the solar system's inner orbits, leaving a concentrated distillate of all the elements in the periodic table.

As soon as the young Earth congealed, its diverse slurry of chemicals generated lumps of rock, solids consisting of many different simple molecules jumbled together. Earth's first minerals also appeared, probably in the form of simple crystals such as graphite or diamonds.[14]

In such a chemically rich environment, many of the simple molecules from which life is built can form more or less spontaneously. We are talking about small molecules, with less than a hundred atoms, including the amino acids from which all proteins are made, the nucleotides from which all genetic material is made, the carbohydrates or sugars that are often used like batteries to store energy, and the fatty phospholipids from which cellular membranes are built. Today, such molecules don't arise spontaneously because atmospheric oxygen would rip them apart. But there was hardly any free oxygen in the atmosphere of the early Earth, so these simple molecules could form when given a few jolts of activation energy.

In 1952, in an effort to demonstrate this, a young University of Chicago chemistry graduate student, Stanley Miller, created a laboratory model of the early Earth's atmosphere by putting water, ammonia, methane, and hydrogen into a closed system of flasks and tubes. He heated the mixture and zapped it with electric charges (laboratory equivalents of volcanoes and electrical storms) to provide some activation energy. Within a few days, Miller found a pinkish sludge of amino acids. We now know that other simple organic molecules, including phospholipids, can also form in such environments. Today, Miller's basic results still stand, even though we know that the early atmosphere was dominated not by methane and hydrogen but by water vapor, carbon dioxide, and nitrogen.

Since then, we have learned that many of these molecules can form even in the less chemistry-friendly environments of interstellar space, so lots of simple organic molecules may have arrived on Earth, ready-made, inside comets or asteroids. For example, the Murchison meteorite, which fell to Earth near Murchison, Australia, in 1969, contained amino acids and several of the chemical bases that we find in DNA. Such meteorites were much more common early in Earth's history than they are today, which suggests that the early Earth was already seeded with many of the raw materials of life and quite capable of manufacturing more.

But most molecules inside cells, such as proteins or nucleic acids, are much more complex than these simple molecules. They consist of polymers, long, delicate chains of molecules, and forming polymers is not so easy. You need just the right amount of activation energy, and environments that can nudge molecules together in just the right way. One environment on the early Earth that might have provided the right conditions for stringing polymers together can be found at suboceanic vents, where hot material from Earth's innards oozes through the ocean floor. These environments were protected from solar radiation and from the violent bombardments on the surface. They also contained diverse chemical elements, lots of water, and gradients of heat and acidity, as hot, chemically rich magmas seeped into cold oceanic waters. Alkaline vents, which were discovered only recently, in 2000, provide particularly promising environments, and the porous rocks that form at these vents offer tiny protected refuges for chemical experimentation, like Miller's flasks and tubes. You can even find claylike surfaces with regular molecular structures that can create physical or electrical templates on which atoms can be wrangled into regular patterns and held still until they form polymerlike chains.

From Rich Chemistry to Life: Luca, the Last Universal Common Ancestor

Life appeared early in the history of planet Earth, and that suggests that creating simple forms of life may not be too hard where the right Goldilocks conditions exist. But identifying exactly when life appeared is tricky because the first organisms lived more than three billion years ago, because they were microscopic, and because the rocks they were buried in have eroded away. At present, the best direct evidence for the earliest life on Earth consists of microscopic fossils found in Western Australia's remote Pilbara region in 2012. They seem to be of bacteria that lived about 3.4 billion years ago.[15] In September 2016, an article in *Nature* described 3.7-billion-year-old fossils of what looked like coral-like stromatolites that were found in Greenland.[16] If these are what many think they are, life must have begun evolving millions of years earlier than previously believed and must have appeared soon after the end of the Late Heavy Bombardment, about 3.8 billion years ago. And early in 2017, on the basis of fossil formations discovered in northern Quebec, scientists claimed that life might have appeared as early as 4.2 billion years ago; we will have to wait to see if these claims stand up.[17]

Biologists don't yet have a complete explanation for how the first living organisms evolved. But they understand many steps in the process.

Though they don't know exactly what it looked like, biologists refer to the first living organism as Luca (or LUCA, from "last universal common ancestor"). Luca certainly lived earlier than the earliest life-forms we have discovered so far, and it shared many features with the modern organisms known as prokaryotes: single-celled organisms whose genetic material is not protected within a nucleus. Today, prokaryotes are found in two of the three large domains of organisms, Eubacteria and Archaea. (The third domain, of which our species is a member, is the Eukarya.)

We'll never find fossils of Luca because Luca is really a hypothetical creature, a sort of composite picture of the first living organism, a bit like a police sketch of a criminal on the run. Still, such a portrait might help us understand how life began.

Luca might have been sort of alive but not fully, somewhere in the zombie zone between life and nonlife. This is not as evasive an idea as it might seem. Viruses are not fully alive because they don't tick all the boxes in our definition of life. They have no metabolism, and they have extremely fragile membranes, so it's not even clear that we can describe them as cells. They are little more than packets of genetic material that glom on to more complex organisms. They enter another cell, hijack the cell's metabolic mechanisms, and use it to make copies of themselves. When you have the flu, viruses are siphoning energy from your metabolic pipelines. But when they can't find cells to hijack, viruses shut down and lurk in a sort of suspended animation. Some cells live deep inside rocks and have extremely slow metabolisms; they survive on tiny scraps of water and nutrition. They may be able to shut down entirely for long periods, like the rock guitarist Hotblack Desiato, in Douglas Adams's novel *The Restaurant at the End of the Universe,* who spends a year dead, for tax purposes. The tax these organisms avoid, of course, is entropy's complexity tax. Luca might have lived in a similar twilight zone.

Composite sketches of Luca have been built up by identifying several hundred genes that are present in most modern prokaryotes and are probably extremely ancient. They suggest the type of environment Luca evolved in, because they tell us what sort of proteins Luca was manufacturing in order to survive.[18]

The composite Luca (or family of Lucas, because we're really talking about billions of them) could adjust to changes in its environment. It had a genome, so it could reproduce. And it evolved. Luca may have lacked both its own membrane and its own metabolism. Its cell walls were probably made of porous volcanic rock, and its metabolism depended on geochemical flows of energy over which it had little control. The proteins Luca made suggest that it lived at the edge of alkaline oceanic vents, probably inside tiny pores in lavalike rocks, and it got its energy from nearby gradients of heat, acidity, and flows of protons and electrons. Luca's chemical innards probably sloshed around in warm liquids from inside the Earth that were alkaline, which meant they had an excess of electrons. Just outside the volcanic pores Luca called home were cooler ocean waters that were more acidic, which meant they had an excess of protons. Like a charged battery, the tiny electrical gradient between Luca's insides and the outer world provided the free energy needed to drive its metabolism, draw in nutrients from outside, and expel waste materials.

Here is how one of the pioneers of early life studies, Nick Lane, describes Luca:

> She [Luca] was not a free-living cell but a rocky laby-
> rinth of mineral cells, lined with catalytic walls com-
> posed of iron, sulphur and nickel, and energised by
> natural proton gradients. The first life was a porous rock
> that generated complex molecules and energy, right up
> to the formation of proteins and DNA itself.[19]

Though simple by comparison with modern organisms, Luca
already contained a lot of neat biochemical gadgets, including
many of the recipes for the metabolic and reproductive machinery of modern cells. It probably had a genome based on RNA so
it could reproduce much more accurately and precisely than
mere chemicals, and that suggests it may have been evolving
fast. It was also using the energy flows it tapped to make ATP
(adenosine triphosphate), the same molecule that transports
energy inside modern cells.

From Luca to Prokaryotes

Luca and its relatives had already done a lot of the heavy lifting
needed to evolve the first true living organisms. But Luca lacked
a membrane that it could carry wherever it went, and a metabolism that was not tethered to energy flows near volcanic vents.
Luca also seems to have lacked the more sophisticated reproductive mechanism that is present in most modern organisms
and is based on RNA's close relative, the double helix of DNA.
At present, we know what had to evolve, but we do not understand the precise pathways by which these things evolved.

Explaining the evolution of personal protective membranes
is not too difficult. Cell membranes are made from long chains
of phospholipids, and it is not hard to persuade phospholipids
to link up in layers that form semipermeable bubblelike structures under the right conditions. Perhaps, as Terrence Deacon

has argued, autocatalytic reactions evolved and generated phospholipid layers, molecule by molecule. If so, it may not be too fanciful to imagine some version of Luca knitting itself a personal membrane.[20]

Explaining how cells evolved better ways of getting energy and reproducing is trickier, but the mechanisms involved are so fundamental and so elegant that it is worth trying to understand how they work.

Evolving new ways of tapping energy flows so that cells could move away from volcanic vents meant creating the cellular equivalent of an electricity grid that molecules could plug into as they went about their work. Enzymes played a crucial role here. These are specialist molecules that can act as catalysts, speeding up cellular reactions and reducing the activation energy needed to get them going. Today, enzymes play a fundamental role in all cells. Most enzymes are proteins, made from long chains of amino acids. The exact sequence of amino acids matters, because that determines how the protein will fold up into the precise shape it needs to do its particular job. Enzymes cruise through the molecular sludge, looking for target molecules that they fit on to, the way a wrench fits a particular nut or bolt. Then the enzyme uses tiny shots of energy to tap, bend, crack, or split the molecule, or bind it to other molecules. Most reactions in your body could not happen without enzymes or would require activation energies so high they would damage the cell.

Once the enzyme has knocked its target molecule into shape, it breaks away and goes hunting for other molecules that it can bend to its will. Enzymes can also be switched on or off by other molecules that bind to them and slightly alter their shape, and this is how, like billions of transistors in a computer, enzymes govern the fantastically complex reactions that go on inside cells.

Enzymes get the energy they need to do their work from the

cellular equivalent of the electrical grid. This is a system that must have evolved very early in the history of life. Energy is carried to enzymes and other parts of the cell by molecules of ATP, or adenosine triphosphate, and ATP was probably hard at work already inside Luca. Enzymes and other molecules tap ATP's energy by breaking off a small group of atoms, releasing the energy that binds that group to the molecule. The depleted molecule (now called ADP, for adenosine diphosphate) then heads off to special generator molecules that recharge it by replacing the lost atoms. The generator molecules are powered by a remarkable process called *chemiosmosis,* which was discovered only in the 1960s but seems to have been at work since the time of Luca. Inside each cell, food molecules are broken down to capture the energy they contain, and some of that energy is used to pump individual protons from inside the cell (where there is a low concentration of protons) to outside the cell (where there is a high concentration of protons). This is like charging a battery. It creates an electrical gradient between the outside and inside of the cell, with a voltage similar to what Luca may have used at alkaline vents. Special generator molecules (ATP synthase, for the technically minded) that are embedded in cell membranes use the electrical voltage created by protons returning from outside the membrane to drive nano-rotors. Like rotary assembly lines, the rotors charge up ADP molecules by replacing the group of molecules they have lost, then the charged-up ATP molecules go back into the cell and wait for other molecules to plug into them and get the energy they need to keep working.

This elegant cellular electrical grid is present in all cells today. It untethered cells from the energy flows around volcanic vents, allowing the earliest prokaryotes to roam Earth's oceans, scrounging energy from food molecules and using them to create ATP molecules that could supply the energy needed to power the cell's innards.

These delicate flows of energy maintained the complex

inner structures of cells just as fusion maintains the structures of stars. Like fusion, they allowed the first living cells to pay the complexity taxes demanded by entropy, because in cells, as in stars, a lot of energy goes into keeping complex structures functioning. But also as in stars, a lot of energy is wasted because no reactions are 100 percent efficient, and of course, entropy loves wasted energy. In both cells and stars, concentrated flows of energy are needed to pay entropy's taxes and overcome the universal tendency of all things to degrade.

In living organisms, energy has a new function that we don't find in stars: it creates copies of the cell. These copies allow cells to push back against entropy by preserving their complex structures even after individual cells have died. Luca's descendants evolved the elegant and efficient methods of reproduction that all living things still use today. Those methods are built on a key molecule, DNA, whose structure was first described in 1953 by Francis Crick and James Watson based on earlier research done by Rosalind Franklin. So much of evolution depends on understanding how DNA works that it is worth looking more carefully at this marvelous molecule.

DNA (deoxyribonucleic acid) is closely related to RNA (ribonucleic acid). Both are polymers, long chains of similar molecules. But while proteins are made from strings of amino acids, and membranes are made from phospholipids, DNA and RNA are made from long strings of nucleotides. These are sugar molecules to which are attached small groups of molecules known as bases. The bases come in four types: adenine (A), cytosine (C), guanine (G), and thymine (T). (In RNA, thymine is replaced by uracil, U.) And here's the magic. As Crick and Watson showed, these four bases can be used like the letters of an alphabet to carry huge amounts of information. As DNA or RNA molecules link up to form huge chains, the bases stick out to the side, forming a long string of As, Cs, Gs, and Ts (or Us in RNA). Every group of three letters codes for a particular amino acid or contains an instruction, such as *Stop reading now*. Thus, the sequence TTA says, *Add on a molecule of the amino acid leucine*, while TAG is a sort of punctuation mark that says, *Okay, you can stop copying now*.

The information on DNA and RNA molecules can be read and copied because the bases like to link up with each other using hydrogen bonds, which can be made and broken quite easily. But they bond only in very specific ways. A always joins with T (or U in RNA), and C with G. Special enzymes expose stretches of DNA that correspond to a particular gene or code for a particular protein, and each base attracts its opposite to create a new short RNA chain of nucleotides that is complementary to the original chain. The newly created segment is then whisked off to a large molecule known as a ribosome, which is a sort of protein factory. The ribosome reads the sequence of letters in triplets and extrudes the corresponding amino acids, one by one, in just the right order to make a particular protein, which then goes off into the cell to do its work. In this way, ribosomes can manufacture all of the thousands of proteins a cell needs.

The final piece of magic is that DNA and RNA molecules can use these copying mechanisms to make copies of themselves and all the information they contain. The bases that stick out sideways from their sugar-phosphate chains reach into the cellular sludge and grab onto their complements. Thus, Cs always grab onto Gs, and As always grab Ts (or Us, in RNA). The newly attached bases attract new sugar molecules that link together, and in this way they form a new chain that is the exact complement of the first. In DNA, these two complementary chains normally stick together, which is why DNA usually exists in the form of a double chain or helix, like a pair of winding staircases. It can be wound up so tightly that it packs neatly inside each cell, and it is unwound only to be read or to make copies of itself. However, RNA normally exists as a single chain, so, like a protein, it can also fold up into particular shapes and function like an enzyme.

This small difference between RNA and DNA is hugely important because it means that, while DNA normally functions just as a store of genetic information, RNA can both store information and do chemical work. It is both hardware and software, and that is why most researchers believe that there was a time, perhaps when Luca was still around, when most genetic information was carried by RNA. Luca probably lived in such an RNA world. But RNA is a less secure information carrier than DNA because its information is constantly buffeted in the violent inner world of the cell, whereas the double strands of DNA shield their precious information from the whirlwind outside. In an RNA world, genetic information could easily get lost or distorted. Evolution really got going only after the development of a DNA world by Luca's descendants, the true prokaryotes, which dominate the world of microorganisms today.

With membranes of their own, an independent metabolism, and more precise and stable genetic machinery, the first pro-karyotes could leave the volcanic vents in which they had been born and cruise the oceans of the early Earth. They were prob-ably already doing this 3.8 billion years ago.

Each prokaryote is an entire kingdom of staggering com-plexity. Billions of molecules swim through a thick chemical slurry, being nudged and pulled by other molecules thousands of times each second, rather like a tourist in a crowded market full of traders, touts, and pickpockets. If you were injected into one of these molecules, you would find this a terrifying world. Enzymes will try to glom on to you and change you, perhaps hook you up with other molecules to form a new team that can cruise the markets looking for new opportunities. Imagine mil-lions of these interactions going on inside every cell every sec-ond and you have some idea of the frenetic activity that powers even the simplest of cells in the early biosphere.

This is a new world and a new kind of complexity. Like stars and planets formed during periods of chaotic change, cells eventually settled into a sort of stability as they began to manage and push back against tiny fluctuations in their environments. Cells would achieve a temporary balance; so, too, would entire species and lineages and groups of species. But this was never a static balance. It was always dynamic, always maintained by a constant negotiation between living organisms and changing environments, and always in danger of a sudden breakdown.

CHAPTER 5

Little Life and the Biosphere

To give Estha and Rahel a sense of Historical Perspective... Chacko... told them about the Earth Woman. He made them imagine that the earth — four thousand six hundred million years old — was a forty-six-year-old woman.... It had taken the whole of the Earth Woman's life for the earth to become what it was. For the oceans to part. For the mountains to rise. The Earth Woman was eleven years old, Chacko said, when the first single-celled organisms appeared.

— ARUNDHATI ROY, *THE GOD OF SMALL THINGS*

Together, Earth and life make up the biosphere.[1] The word *biosphere* was coined by the Austrian geologist Eduard Suess (1831–1914). Suess saw Earth as a series of overlapping and sometimes interpenetrating spheres that included the atmosphere (the sphere of air), the hydrosphere (the sphere of water), and the lithosphere (the rigid, upper levels of the Earth, including the crust and the top layers of the mantle). But it was the Russian geologist Vladimir Vernadsky (1863–1945) who first showed that the sphere of life has shaped planetary history as powerfully as the other, nonliving spheres. We can think of the biosphere as a thin wrapping of living tissue (and the remains and imprints of living tissue) that reaches from the depths of the oceans to Earth's surface and up into the lower atmosphere. In the 1970s, James Lovelock and Lynn Margulis showed that the biosphere can be thought of as a system with many feedback mechanisms that allow it to stabilize itself in the absence of major shocks. Lovelock called this vast, self-regulating system Gaia, after the Greek goddess of the Earth.

Geology: How Planet Earth Works

Life took some time to get going, so we will begin by considering planet Earth as a purely geological system, like a stage set before the actors have arrived. That should make it easier to understand the complex dramas acted out later by living organisms.

The violent processes of accretion and differentiation, which had forged the young Earth, left a chemically rich ball of matter separated into distinct layers. There was a hot, semimolten core, made mostly of iron and nickel, that generated a protective magnetic field around Earth. Wrapped around the core was a three-thousand-kilometer-thick layer of gas, water, and semi-molten rock, the mantle. The lightest rocks rose to the surface and formed Earth's crust. Gases and water vapor bubbled up through volcanoes to create Earth's first atmosphere and oceans. Meteors and asteroids ferried in new cargoes of rocks, minerals, water, gases, and organic molecules.

About 3.8 billion years ago, when the bombardment from space eased up, the main driver of geological change was the heat buried in Earth's core. That heat seeped up through Earth's mantle, to the crust, and into the atmosphere, churning up the material in each layer, transforming it chemically, and moving vast amounts of matter and gas around in huge, slow convection cycles. Like the evolution of stars, the geological evolution of our Earth was driven primarily by simple processes that fed on an initial, nonrenewable store of energy. Earth changed as it sweated heat from the core through the mantle and crust and out into space.

Heat from the core still drives a lot of geology and will continue to do so for billions of years. But not until the 1960s did geologists figure out how this huge geological machine worked. Their new understanding of geology was based on one of modern science's most important paradigms: plate tectonics.

Humans have been able to visualize Earth's surface only in the past five hundred years, when, for the first time, they were able to sail all around it. But most people continued to assume that at large scales, the world's geography was more or less fixed. Volcanoes might erupt and rivers change course, but surely the layout of continents and oceans, of mountains, rivers, and deserts, of ice caps and canyons, was unchanging. Some, though, began to have doubts. And, just as Darwin showed that life had changed profoundly over the eons, evidence began to accumulate that Earth, too, had a history of profound change.

In 1885, Eduard Suess suggested that about two hundred million years ago, all the continents had been joined together in one supercontinent. We now know he was dead right. Three decades later, Alfred Wegener, a German meteorologist who had done research in Greenland, assembled a lot of evidence that supported Suess's idea. Wegener published that evidence in 1915, the middle of World War I, in a book entitled (perhaps with a nod to Darwin's *Origin of Species*) *The Origin of Continents and Oceans.* Just as Darwin proposed that living organisms had evolved, Wegener proposed that continents and oceans had evolved, by a mechanism he called continental drift. Once joined in the supercontinent of Pangaea, or Pan-Gaia (a Greek word meaning "all Earth"), they had gradually diverged and moved to their present positions.

Wegener offered plenty of evidence. On a world map, many parts *look* as if they once fit together, something people had noted since the creation of the first world maps in the sixteenth century. Just before 1600, a Dutch mapmaker, Abraham Ortelius, commented that the Americas seemed to have been "torn away" from Europe by some catastrophe.[2] If you look at a modern world map, you'll see that the shoulder of Brazil fits nicely into the armpit of western and central Africa, while West Africa looks as if it would fit snugly into the huge arc of the Caribbean. In the 1960s, geologists realized that the fit is even better if you focus on the edges of the continental shelves.

Wegener showed that there were almost identical fossils of ancient reptiles in South America and central and South Africa. The early nineteenth-century German scientist Alexander von Humboldt, one of the first scholars to write a modern, science-based origin story, had also noticed similarities between the coastal plants of South America and Africa.[3] Then there were rock strata that seemed to start in West Africa and continue in eastern Brazil without missing a beat. As a meteorologist, Wegener was particularly interested in climatic evidence. In tropical Africa, you could find the telltale scratches and gouges of moving glaciers. Could tropical Africa once have hovered over the South Pole? In Greenland, Wegener had found fossils of tropical plants. Something had certainly moved over long distances in the deep past.

But it takes more than some suggestive evidence to make a good scientific hypothesis. Publishing in the middle of World War I didn't help Wegener's case, and the fact that he was German and *not* a geologist ensured that few geologists in the English-speaking world took his ideas seriously. Was it *really* possible that whole continents could plow through the oceans? Wegener had no idea what force could have pushed them around, and in the eyes of most professional geologists, the absence of an explanation was enough to kill off his hypothesis. In November 1926, Wegener's theory of continental drift was decisively rejected by the influential American Association of Petroleum Geologists. And that seemed to be that.

Except that a few geologists were intrigued. A British geologist, Arthur Holmes, argued in 1928 that the interior of Earth might be hot enough to act like a slowly moving liquid, like lava. If so, perhaps the motion of material inside Earth could float entire continents around the globe. But not until the 1950s would new evidence show that Wegener, Holmes, and other supporters of the idea of continental drift had been following the right geological scent.

That's where sonar (the word comes from "sound navigation ranging") enters the story. Sonar technology can detect and locate objects underwater by bouncing signals off them and analyzing the returning echoes. Many animals use sonar, including dolphins and bats. Human sonar technology, like radiometric dating, was a product of wartime science, in this case attempts to detect enemy submarines. Harry Hess, a geology professor at Princeton, was a naval commander during World War II, and he had used sonar to track German submarines. After the war, he used sonar to map the seafloor, which was still unknown territory to marine geologists. Most expected the seafloor to consist of a flat ooze washed off the continents. Instead, Hess found chains of volcanic mountains running through the Pacific Ocean. No geologist had expected that. After discovering a similar chain running through the middle of the Atlantic Ocean in the early 1950s, he began to develop a theory to explain these mid-oceanic ridges. His task was helped by paleomagnetism, or studies of the magnetism of the seafloor. It was already known that at intervals of up to a few hundred thousand years, Earth's north and south magnetic poles had swapped places many times. These flips left their traces in lava that seeped up through the ocean floor and took on the prevailing magnetic orientation as they solidified. Measurements of the magnetic orientation of rocks on either side of the volcanic ridges showed a series of north/south flips as you moved away from the ridges. This puzzled Hess.

Eventually, Hess figured out that the undersea mountain chains were being created by magma squeezed up through cracks in the oceanic crust. This made sense, because oceanic crust is thinner than continental crust, so it can be punctured easily by hot magma. As magma climbed through cracks in the seafloor crust, it elbowed the crust apart, creating new seafloor that was imprinted with the magnetic orientation of the period when it formed. The alternating magnetism of mid-oceanic rocks provided a way of dating the formation of these underwater mountain ranges.

Lurking in these discoveries lay the driver of continental drift that Wegener had looked for in vain. Mountain chains, continents, and the seafloor were created and pushed around by huge amounts of hot magma that rose from Earth's mantle and squeezed through cracks in the seafloor crust. The magma was heated by radioactive elements and by heat from Earth's core, which retained much of the energy stored during the violent processes of accretion and Earth building. And there in the planet's core lay the missing driver. Like fusion at the center of a star, heat leaking from the center of the Earth drives most important geological processes on the surface.

We now have abundant evidence that Earth's crust, both oceanic and continental, is broken into distinct plates that jostle for position as they are dragged back and forth by the semi-molten magma on which they float. Hot magmas rising from deep within the Earth circulate under the crust, like water boiling in a saucepan. It is these convection currents of semiliquid rock and lava that move the tectonic plates floating above them. Detailed studies of paleomagnetic bands have allowed earth scientists to trace the movements of plates over hundreds of millions of years, giving us an increasingly precise idea of Earth's changing geography over the last billion years or so. We now know that these movements have created supercontinents like Pangaea and then broken them up several times in a cyclic process that probably began early in the Proterozoic eon, about two and a half billion years ago. Before that, there were probably no large continents. But some geologists argue that the machinery of plate tectonics may have powered up much earlier. Evidence from the Hadean eon suggests that some form of plate tectonics was already at work 4.4 billion years ago, as soon as Earth differentiated into distinct layers.[4]

Like big bang cosmology, plate tectonics was a powerful unifying idea. It explained and showed the links between many different processes, from earthquakes to mountain building and the movement of continents. It explains why so many violent geological events take place where tectonic plates meet and grind their way past, over, and under each other. Plate tectonics also explains why Earth's surface is so dynamic, as it is continually renewed by the arrival of new materials from the mantle, while surface material, in turn, is subducted deep into the Earth.

To understand how plate tectonics works in detail, it helps to focus on the borders between tectonic plates. At *divergent margins,* like those described by Harry Hess, material rises from the mantle and pushes plates apart. Elsewhere, though, at *convergent margins,* plates are pushed together. If the two plates have about the same density—if, say, they both consist of granitic continental plates—then, like two bull walruses competing for mates, they will rear up. This is how the Himalayas formed; within the past fifty million years, the fast-moving Indian plate traveled north from Antarctica and smashed into the Asian plate. But if two converging plates have different densities—if, say, one consists of heavy, basaltic oceanic crusts and the other of lighter granitic continental crust—the story is different. The heavier oceanic plate will dive under the lighter plate at a subduction zone. It will travel downward, like a runaway elevator crashing through a concrete floor, carrying crustal material back into the mantle, where it dissolves. As the descending plate drills into the mantle, it will generate so much friction and heat that it can melt the crust above it, splitting it and punching up new volcanic mountain chains. This is how the Andes formed, as the Pacific plate burrowed beneath the plate carrying the west coast of South America.

Finally, there are *transform margins*. Here, plates grind their way past each other like two pieces of sandpaper jammed together but pushed in opposite directions. Friction will stop the plates sliding until so much pressure builds up that there is a sudden, violent lurch. This is the source of the pressure building up along the San Andreas Fault on the western coast of North America. (Living for a while in San Diego, I occasionally felt tremors, and, like many Californians, I had to buy earthquake insurance.)

The circulation of materials between the atmosphere, the surface, and the mantle had a profound impact on the chemistry of Earth's upper layers. It generated new types of rocks and minerals. By the time life began to flourish on land, chemical processes within the mantle had already created as many as fifteen hundred distinct types of minerals.[5] Plate tectonics give planet Earth an exceptional chemical and geological dynamism.

Plate tectonics also affected temperatures at the young Earth's surface, and we have already seen how crucial the right temperature was to the history of life on Earth. Two major forces determine average temperatures at Earth's surface: heat from the interior and heat from the sun. These we can roughly calculate. But the composition of the atmosphere helps determine how much heat is retained at Earth's surface and how much leaks away into space. Particularly important is the proportion of greenhouse gases. These are gases like carbon dioxide and methane that trap the energy of sunlight rather than reflecting it back into space. Large amounts of greenhouse gases generally mean a warmer Earth. So what controls levels of greenhouse gases?

The astronomer Carl Sagan (one of the great pioneers of a modern origin story) pointed out that answering this question is vital because it may solve another puzzle. Stars like our sun emit more and more energy as they age, so the amount of heat arriving on Earth has slowly increased. When Earth was young, the sun was emitting 30 percent less energy than today. So why was the early Earth not a ball of ice and far too cold for life to form, like Mars today? Carl Sagan called this problem the "early faint sun paradox."

The answer, it turned out, was the amount of greenhouse gases in the early atmosphere. Their levels were high enough to warm the young Earth so that life could evolve. There was hardly any free oxygen in Earth's first atmosphere, but there were lots of greenhouse gases, particularly water vapor, methane, and carbon dioxide, belched up from the mantle through volcanoes or ferried in by asteroids. A greenhouse atmosphere was one more important Goldilocks condition for life on the young Earth.

But how stable was this early greenhouse atmosphere? Or, to put it more generally, what ensured that as the sun began to emit more energy, Earth's surface would remain within the magical temperature range of zero to one hundred degrees Celsius? In the 1970s, James Lovelock and Lynn Margulis argued that there seemed to be powerful self-regulating mechanisms that kept Earth's surface within the Goldilocks range. As we have seen, they called that something Gaia. Gaia consisted of the sum total of relationships between Earth's geology and its living organisms that was keeping Earth life-friendly. Many scientists remain skeptical of the Gaia hypothesis. But what is clear is that there are indeed feedback mechanisms within the biosphere, and many do act like thermostats to partially regulate the temperature at Earth's surface. Some mechanisms are geological, but others work through living organisms.

One of the most important of these thermostats is purely geological, so it would have begun to work even before there was life on Earth. It links tectonics and another driver of planetary change: erosion. While tectonics builds mountains up, erosion grinds them down. Wind and water and chemical flows of various kinds break down the rocks of mountains and move them down a gravitational gradient into the oceans. Erosion explains why mountains aren't much higher than they are; tectonics explains why they haven't all vanished into a single, vast global plain. Erosion is itself a by-product of tectonics, of course, because both the wind and rain were burped up from Earth's innards. And mountain building can speed up erosion as gravity turns high mountain rivers into destructive torrents that gouge the land and transport soils fast toward the ocean.

Here's how the geological thermostat works. Carbon dioxide, one of the most powerful of the greenhouse gases, dissolves in rainfall and reaches the Earth in the form of carbonic acid. It dissolves material in rocks, and the by-products of these reactions, which contain lots of carbon, are swept into the ocean. Here, some of the carbon gets locked up in carbonate rocks. Where tectonic plates dive back into the mantle at subduction zones, some of this carbon (much of it in the form of limestone) can get buried in the mantle for millions, even billions of years. In this way, the tectonic conveyor belt removes carbon from the atmosphere, and that should eventually reduce carbon dioxide levels and generate colder climates. Today we know that much more carbon is buried within the mantle than is present on Earth's surface or in its atmosphere.

Of course, if too much carbon dioxide was buried in this way, Earth would freeze. That was prevented (most of the time) by the second feature of the geological thermostat. Driven by plate tectonics (a mechanism that is probably not working on icy Mars), carbon dioxide can return to the atmosphere at divergent zones, where material from the mantle, including buried carbon dioxide, rises to the surface through volcanoes.[6] There is a balance between the two halves of this mechanism because higher temperatures generate more rainfall, which accelerates erosion, moving more carbon back into the mantle. But if the Earth cools too much, rainfall will dwindle, less carbon dioxide will be buried, carbon dioxide levels will build up as it is pumped up through volcanoes, and that will warm things up again. The geological thermostat has been adjusting to the increasing warmth of our sun over four billion years.[7]

We know of nothing like this happening on other planets in our solar system. Venus suggests what Earth could have been like if too much carbon dioxide had remained in the atmosphere. Today, Venus's atmosphere contains huge amounts of carbon dioxide, and the planet seems to have suffered from a runaway greenhouse effect. Its surface is hot enough to vaporize water and melt lead. Mars took the other wrong turn. It was too small for its gravity to hold on to greenhouse gases, so they leaked away; the planet cooled, and most of its water now exists in the form of ice. Curiosity Rover, as it crawls across the surface of Mars, has shown that there was a time, billions of years ago, when water flowed across its surface and simple life-forms might have flourished. But that time is long past. In any case, neither Mars nor Venus seem to do plate tectonics, which deprives them of a key component of our planet's thermostat. Mars was too small to retain the internal heat needed to drive tectonics, and Venus, by boiling most of its water away, may have deprived tectonics of the watery lubricant that helped plates move past and over and under each other.[8]

The geological thermostat was far from perfect, and there were times when it threatened to break down, which would have had dire consequences for the biosphere. But eventually, other, backup thermostats evolved. These were created by the activities of living organisms. So now we must return to the role of life in the biosphere as living organisms stepped onto Earth's geological stage and began to explore and eventually transform its many different ecological nooks and crannies.

The Unity of Life

Despite the profound differences between *Tyrannosaurus rex* and an *E. coli* bacterium, in important respects, life is remarkably unified. All organisms alive today are related genetically. And they share many genetic gadgets, particularly those that, like subroutines in computer software, handle basic housekeeping tasks. In cells, these tasks include jobs such as breaking down food molecules for their energy or their chemical components or moving energy and atoms around. This is why, if you zoom down to the level of cells, it's hard to distinguish between a human being and an amoeba.

Today, biologists can track the genetic relationships among all living organisms by comparing the huge sequences of As, Cs, Gs, and Ts in their DNA. The basic rule is that the more divergence there is between two genomes, the longer it's been since those two species shared a common ancestor, and we know roughly the speed at which different types of genomes diversify. So we can say with some confidence that humans and chimps shared a common ancestor about seven or eight million years ago, while humans and bananas have followed different genetic paths for about eight hundred million years. Comparing the DNA of different living species allows us to construct family trees that are much more detailed, and probably more precise, than those based just on the fossil record.

Today, biologists classify all living organisms into three great domains: Archaea and Eubacteria, which consist entirely of single-celled prokaryotes, and Eukarya, which consists of more complex single-celled organisms and also multicelled organisms such as ourselves. The modern classification system has evolved from the taxonomic (classificatory) work of the eighteenth-century Swedish biologist Carl Linnaeus. He grouped all organisms into nested classes. The lowest taxonomic level, the species, contains just one entry. The next level up is the genus, a group of closely related species. Humans, for example, belong to the genus and species *Homo sapiens;* the genus *Homo* includes our now-extinct ancestors *Homo habilis* and *Homo erectus* (also known as *Homo ergaster*). The taxonomic levels become increasingly capacious from there; in ascending order, they are family, order, class, phylum, kingdom, and domain. So we can say that humans belong to the species *sapiens,* the genus *Homo,* the family Hominidae, the order Primates, the class Mammalia, the phylum Chordata (vertebrates), the kingdom Animalia, and the domain Eukarya.

The first living organisms surely diversified fast, as they entered new evolutionary territory. Many zombies may have survived among them. Here's one description of the strange world of early life from a recent history of life on Earth:

> We can think of a giant zoo of the living, the near living, and the evolving towards living. What would that zoo contain? Lots of nucleic acid creatures of many kinds, things no longer existing and having no name because of this. We can imagine complicated chemical amalgamations. And all these huge menageries of the living and near living would have existed in one thriving, messy, competing ecosystem — *the time of life's greatest diversity on Earth.*[9]

Sometime early in the Archean eon (which started four billion years ago), reproductive mechanisms got more precise, genes got more stable, and the borders between the living and the almost-living got clearer. That is the point at which natural selection, in Darwin's sense, really took off. Once life got going, there were no guarantees it would survive. Mars and Venus may once have hosted simple life-forms. But if they did, life was soon extinguished on both planets. Even on Earth, the survival of a thin scum of life for almost four billion years depended on lots of things going right.

Prokaryotes: A World of Single-Celled Organisms

The first living organisms probably belonged to the domain of Archaea, though organisms from the second domain, the Eubacteria, also appeared early. Both domains consist entirely of prokaryotes, minute single-celled organisms that have neither a distinct nucleus nor other specialized cellular organelles. Prokaryotes would dominate the biosphere for more than seven-eighths of its history, until about six hundred million years ago. If we meet living organisms elsewhere in our galaxy, we probably won't be shaking hands with them but peering at them through a microscope.

So small are prokaryotes that one hundred thousand of them could have a party inside the dot at the end of this sentence. Prokaryote genes float freely in rings and filaments inside the salty molecular sludge of their cytoplasm, so their DNA is constantly buffeted, like everything else in the cytoplasm, and can easily be damaged or altered. Bits of genetic material could even float through the cell membranes and migrate to other cells. In the prokaryotic world, many genetic ideas spread sideways, among unrelated individuals, as well as vertically, from parent to offspring. Prokaryotes trade genes as we humans trade stocks and shares, which is why the idea of a distinct species is harder to define in the prokaryotic world than in our world.

Today, prokaryotes still dominate the biosphere. On and

within your body, there are probably more prokaryotic cells than cells with your own DNA. But we ignore them (until they give us a stomachache or cold) because they are so much smaller than our own cells. This vast shadow world that we share with prokaryotes is known as the *microbiome.*

Until recently, it was tempting to think that the history of single-celled organisms was boring, so we could happily skip the first three billion years of the biosphere's history. Today we are learning that we can't make sense of the biosphere's recent history without understanding the much longer era of little life. As they evolved, prokaryotes developed many new tricks that let them exploit different environments, and we still use several of the biochemical techniques they pioneered.

All prokaryotes can process information. In a sense, they can even learn. Embedded in their membranes are thousands of molecular sensors that can detect gradients of light and acidity, sense when there are potential foods or poisons nearby, and tell if they have bumped into something hard. The sensors are made of proteins, which, like all enzymes, have binding sites that glom on to particular molecules outside the cell or react to changes in light, acidity, or temperature. Once these proteins detect something, their shape changes slightly, which sends a signal to the inside of the cell. The much-studied *E. coli* bacterium, for example, has four different types of sensor molecules embedded in its membranes, and together they can detect about fifty different types of good or bad things in the neighborhood.[10] Once the sensors have detected something, the cell can make choices. For example, it can decide to let particular molecules through its membrane walls (because they look like food) or keep them out (because they look like poison). The decision-making can be very simple. It may be based on a tiny number of inputs and require only yes/no answers. "Should I let this molecule in or not?" Or "It's getting too hot on this side, yikes! Should I move?" But even the simplest sensors are, in effect, creating basic sketches of a cell's environment. Once a decision to move has been reached, any equipment the cell has to control motion will be activated. For many bacteria, that is a sort of rotating tentacle, or flagellum, that can act like a propeller. *E. coli* has six of these whiplike appendages embedded in its membranes. Each is constructed from twenty different components and can rotate several hundred times a second, powered by energy from proton gradients across its membranes. When needed, the flagella can rotate together to give a more directed motion.[11] The link between sensors in the membrane and the flagella means that, in effect, *E. coli* has a short-term memory. It may last for just a few seconds but is powerful enough to say either "No problem, nothing to do!" or "This is not good, flagella, *start flailing!*" The short-term memory is based on tiny changes in the sensors and the chemicals they emit.

This is simple information-processing equipment, but already we have the three key components of all biological information processing: inputs, processing, and outputs.

Information management gave prokaryotes more control over local flows of energy. Over time, prokaryotes evolved to get, control, and manage energy in many of the diverse environments of Earth's oceans. The first prokaryotes were probably chemotrophs. That means they got their energy from geochemical reactions between water and rocks that released simple substances such as hydrogen sulfide and methane, chemical energy they could tap.[12] But easily digestible chemicals that could release drip-feeds of energy were in limited supply in the earliest oceans; they were readily available only in rare environments such as suboceanic vents. Those limits would have narrowed the possibilities for life on Earth. Quite early on, some prokaryotes learned how to eat other prokaryotes. These were the biosphere's first heterotrophs, the prokaryotic equivalent of carnivores such as *T. rex*. You and I are also heterotrophs; we get our food energy by consuming other organisms, not by eating chemicals. But even eating other organisms has its limits if the entire biosphere depends on an energy chain anchored within the oceans.

Photosynthesis: An Energy Bonanza and a Revolution

By about 3.5 billion years ago, a new evolutionary innovation, photosynthesis, was letting some organisms tap into flows of energy from the sun. This was life's first energy bonanza, and its impact was the prokaryotic equivalent of a gold strike.

Photons of light from the sun have thousands of times more energy than the tired old photons from the cosmic background radiation. Tapping into that colossal flow of energy was a game changer. From now on, though life would continue to recycle all the matter it used (hence the interest of scientists in flows of carbon, nitrogen, and phosphorus), energy seemed to be more or less limitless.[13] Living cells now had the energy to reorganize themselves and their surroundings on an entirely new scale. They spread more widely and the amount of life surely increased by several orders of magnitude.

How did living organisms use sunlight? There are several types of photosynthetic reactions that convert sunlight to biological energy with varying degrees of efficiency and release different by-products. All of them use energetic photons newly arrived from the sun to goose electrons inside light-sensitive molecules such as chlorophyll. This gives the electrons such a shock that they jump out of their home atoms and then get kidnapped, wriggling all the time, by proteins. The proteins pass the high-energy electrons through cell membranes in a sort of bucket brigade. This creates an electrical gradient across the membrane that can be used to charge up energy-carrying molecules such as ATP. This is chemiosmosis again, but this time, the energy that charges up molecules of ATP comes not from food molecules but from that vast generator in the sky, the sun.

That's the first stage in all forms of photosynthesis. In the second stage, the captured energy is used, in a series of complex chemical reactions that vary greatly in their efficiency, to do work inside the cell or to form molecules such as carbohydrates that can warehouse energy for future use. The earliest forms of photosynthesis did not produce oxygen as a by-product, and they worked well in a world without free oxygen. They may have used energy captured from sunlight to steal electrons from hydrogen sulfide (rotten-egg gas) or from iron atoms dissolved in the early oceans.

Even the simplest early forms of photosynthesis provided a revolutionary new supply of energy, and the amount of life in the early oceans may have increased to as much as 10 percent of today's levels.[14] Prokaryotes that made their living from photosynthesis had to be near the surface of the oceans or on seashores. Many formed coral-like structures known as stromatolites, which grew into reefs at the edges of continents as billons of organisms accumulated over thickening layers of their dead ancestors. Stromatolites still exist in a few special environments, such as Shark Bay, off the coast of Western Australia. They are rare today, but from the time when they first appeared, more than 3.5 billion years ago, until about 500 million years ago—significantly more than half the history of our planet—they were probably the most visible form of life on Earth. If aliens had come looking for life on this planet, they would have found stromatolites. And perhaps that's what we'll find when we first detect life on rocky planets in other star systems.

Eventually, new forms of photosynthesis evolved in a group of organisms known as cyanobacteria. These forms of photosynthesis could extract more energy by using water and carbon dioxide as their primary raw materials. Prying electrons loose from water molecules was tougher than capturing them from hydrogen sulfide or iron. But if you could do it, you got more energy, and of course in water, you had a much more abundant source of energy. Using the energy captured from sunlight, these sophisticated photosynthesizers zapped water molecules and stripped electrons from their hydrogen atoms. Then they added the captured electrons to carbon dioxide molecules to form carbohydrate molecules, which acted as huge energy barns. The oxygen from broken water molecules was released as waste. The general formula for this oxygen-generating form of photosynthesis is $H_2O + CO_2$ + energy from sunlight $\rightarrow CH_2O$ (carbohydrates that act as stores of energy) + O_2 (molecules of oxygen that are released into the atmosphere). Oxygen photosynthesis was much more efficient than earlier forms but still could extract only about 5 percent of the energy in sunlight, which is less than the most efficient modern solar panels. Photosynthesis pays a substantial garbage tax to entropy in the energy wasted inside the cell and the energy and materials carried away by oxygen.

Oxygen-producing photosynthesis, the sort of photosynthesis used by all modern cyanobacteria, may have evolved as early as three billion years ago. This is suggested by evidence for brief "whiffs" of increased oxygen levels even before the end of the Archean eon, two and a half billion years ago. But at first, any oxygen they released would have been quickly absorbed by iron or hydrogen sulfide or free hydrogen atoms, because oxygen is an electron thief and will combine eagerly with any element that has spare electrons. That is why atoms that have had their electrons stolen are said to have been oxidized. (Atoms with spare electrons are said to be reduced, and the many chemical reactions that involve both processes are known as redox reactions.) Powerful evidence for the evolution of the first cyanobacteria is the disappearance from three billion years ago of sedimentary rocks rich in pyrite (fool's gold), which, like iron, rusts in the presence of free oxygen. But there was a limit to how much oxygen these mechanisms could absorb, and starting about 2.4 billion years ago, levels of atmospheric oxygen began to rise fast, from less than 0.001 percent of today's levels to perhaps 1 percent or more.

The appearance of an oxygen-rich atmosphere beginning about two and a half billion years ago (the "great oxygenation event") transformed the biosphere. Rising oxygen levels altered the chemistry of the biosphere and even of the upper levels of Earth's crust. The exceptional chemical energy of free oxygen powered new chemical reactions that created many of the minerals on today's Earth.[15] High up in the atmosphere, oxygen atoms combined to form three-atom molecules of ozone, O_3, that began to shield Earth's surface from dangerous solar ultraviolet radiation and have continued to do so ever since. Protected by the ozone layer, some algae may have started colonizing the land for the first time. Until then, bathed in solar radiation that would have ripped apart any bacteria brave enough to venture onto land, the continents of planet Earth had been more or less sterile.

The oxygen buildup was a profound shock to living organisms because, for most of them, oxygen was poisonous. So rising oxygen levels caused what the biologist Lynn Margulis called an "oxygen holocaust." Many prokaryotic organisms perished, and those that didn't die retreated to protected environments in the deeper, oxygen-poor levels of the oceans or even into rocks.

Rising oxygen levels messed up Earth's thermostats because as yet there were no mechanisms that could absorb excess oxygen, so the buildup threatened to run out of control. Free oxygen broke down atmospheric methane, one of the most powerful of greenhouse gases, while photosynthesizing cyanobacteria consumed huge amounts of the other crucial greenhouse gas, carbon dioxide. As oxygen levels rose and levels of greenhouse gases fell, early in the Proterozoic eon, Earth froze in the first of several snowball-Earth episodes. Glaciers spread from the poles to the equator, turning the Earth white, and a white Earth reflected more sunlight, cooling it even further in a terrifying positive-feedback loop. Eventually, most of Earth's oceans and continents were covered by ice. The Makganyene glaciation lasted for one hundred million years, from around 2.35 to 2.22 billion years ago.

This was a close shave. Organisms for which oxygen was a poison perished or hid deep in the oceans. But even organisms that could cope with oxygen suffered in a world where glaciers covered both the land and the oceans, blocking the sunlight needed for photosynthesis. Life hung by a thread, as most life-forms retreated beneath the ice and huddled around the warm fires of suboceanic volcanoes.

But Earth did not go the way of Mars and get too cold for life. This was thanks to the geological thermostat driven by plate tectonics, now renovated and supplemented by new biological techniques that depended on the activity of photosynthesizing organisms. Glaciers blocked photosynthesis, which slashed oxygen production. Meanwhile, under the glaciers, oceanic volcanoes kept pumping carbon dioxide and other greenhouse gases back into the oceans. Greenhouse gases began to accumulate beneath the ice until, eventually, they broke through the glaciers, and Earth's surface warmed again. Oxygen levels plummeted to about 1 or 2 percent of the atmosphere, and there followed a long period, almost a billion years, during which oxygen levels remained low and climates remained warm. Earth's ancient thermostats seemed to have been reset to cope with the presence of significant levels of atmospheric oxygen produced by cyanobacteria.

Eukaryotes to the Rescue

Was this a long-term solution? Didn't these mechanisms promise a biosphere that would fluctuate dangerously between extreme heat and extreme cold? If so, why were climates relatively stable for a billion years between about two billion and one billion years ago? Now biology came to the rescue by evolving new types of organisms that could supplement Earth's thermostats by sucking oxygen out of the air. These organisms, the first eukaryotic cells, didn't just help stabilize global temperatures. They also marked a biological revolution that would eventually allow the evolution of large organisms such as you and me.

So far, all living organisms had been single-celled prokaryotes in the domain of either Archaea or Eubacteria. The appearance of a third domain of life-forms, Eukarya, matters a lot to us because all large organisms, including ourselves, are built from eukaryotic cells. These were the first cells that could use oxygen systematically, exploiting its fierce chemical energy in a process known as respiration, which is what we do when we breathe. Respiration is the reverse of photosynthesis and is really a way of releasing solar energy that has been captured and stored within cells through photosynthesis. While photosynthesis uses energy from sunlight to turn carbon dioxide and water into energy-storing carbohydrates, leaving oxygen as a waste product, respiration uses the chemical energy of oxygen to pilfer the energy warehoused in carbohydrates, leaving carbon dioxide and water as waste products. The general formula for respiration is CH_2O (carbohydrates) $+ O_2 \rightarrow CO_2 + H_2O +$ energy.

Like photosynthesis, the evolution of respiration by eukaryotes counts as an energy bonanza, because it gave these new organisms access to the huge chemical energies of oxygen but in tiny, gentle doses that didn't blast them apart. Respiration gives you the energy of fire without its destructiveness. By using oxygen cleverly, respiration can extract at least ten times as much energy from organic molecules as earlier non-oxygenic ways of breaking down food molecules.[16] With more energy to power their metabolism, rates of primary production—the production of living organisms—may have increased by anything from ten to a thousand times.[17]

Genetic evidence suggests that the first eukaryotes evolved about 1.8 billion years ago.[18] As they proliferated, taking in more and more oxygen, they pumped carbon dioxide back into the atmosphere as a waste product. And here we see the beginnings of a new, biologically controlled planetary thermostat. Eukaryotes began to remove much of the atmospheric oxygen generated by cyanobacteria. This may help explain why climates were relatively stable for much of the Proterozoic eon. Indeed, they were so stable that some paleontologists refer to the period between about two and one billion years ago as "the boring billion."

Modern biologists regard the distinction between eukaryotic and prokaryotic cells as one of the most fundamental divides in biology. Eukaryotic cells are much larger than most prokaryotic cells. They can be ten or a hundred times as wide, so their total volume can be many thousands of times as large. In eukaryotes, membranes form inside cells as well as around them, creating separate compartments, like rooms in a house, in which different activities can take place. This allows specialization, an internal division of labor that was impossible for prokaryotes. One of these compartments, the nucleus, protects the genetic material of all eukaryotes. Indeed, the word *eukaryote* comes from the Greek for a "shell" or "kernel." The protected container of the nucleus ensured that eukaryotic DNA was generally more stable than that of prokaryotes. It could also be stored in larger amounts and copied more easily, so eukaryotes generally have more genetic toys to play with. That explains why they would eventually evolve even more exuberantly than prokaryotes. Eukaryotes also contain many internal organelles, like cut-down versions of the hearts, livers, and brains of animals. The most important of these are the mitochondria that some eukaryotes use to tap the rich energy of oxygen, and the chloro-plasts that other eukaryotes use to tap the energy of sunlight through photosynthesis.

Eukaryotes also had new information-processing and body-control capabilities, which meant they could respond in more complex ways to changes in their surroundings.[19] The single-celled eukaryote paramecium has a neat trick for dealing with obstacles. If it hits one, it backs off, turns a few degrees, and moves forward again, repeating the toing-and-froing, like a bad driver trying to parallel-park, until it is no longer hitting any-thing. In effect, it is mapping its environment and learning what to do next. It is using information about its surroundings to ori-ent itself in the world, to avoid dangers, and to find energy and food.

How did the first eukaryotic cells evolve? The biologist Lynn Margulis showed that they evolved not through competition but rather by a sort of merging of two existing prokaryotic species. It is common for different species to collaborate through what is known as symbiosis. Today, humans have vital symbiotic rela-tionships with wheat, rice, cattle, sheep, and many other species. But Margulis was talking about a much more radical type of symbiosis, one in which once independent bacteria, including the ancestors of modern mitochondria, ended up living *inside* a cell from the Archaea. Margulis called the mechanism *endosym-biosis*. At first, her idea seemed crazy, because it ran counter to some of the most fundamental concepts about evolution by nat-ural selection. But most biologists now accept her arguments.

The most important evidence for endosymbiosis is the odd fact that some of the organelles inside eukaryotes contain their own DNA, and that DNA is quite different from the genetic material in the nucleus. Margulis realized that organelles such as the mitochondria that manage energy in animals and the chloroplasts that manage photosynthesis in eukaryotic plants *look* as if they were once independent prokaryotic cells. Exactly how they ended up inside other cells remains unclear, and some have argued that such mergers must be extremely rare. If so, that probably means that even if bacterialike organisms are common in the universe, large organisms like us may be extremely rare, because, on our planet at least, only eukaryotes can build large organisms.

Margulis's discovery of endosymbiosis tells us something more about the history of life. Evolution is not just a matter of competition. Nor is it just a matter of constant divergence as new species appear. We also see collaboration, symbiosis, and even convergence. This means we have to reconsider the conventional metaphor of a tree of life, because even if we still think of three domains of life, it looks as if the third domain, the Eukarya, evolved not by increasing divergence but by the convergence of Archaea and Eubacteria — rather as if two branches of an ancient tree joined up again.

As if that were not strange enough, eukaryotes had one more trick up their sleeve: sex. Like all species, prokaryotes pass their genes on to their offspring. Most just split in two and pass on their genes through asexual reproduction. But, as we have seen, prokaryotic genes can also travel sideways as bits of DNA and RNA jump ship, go on the road, and find new homes inside other cells. Prokaryotic cells share genes the way humans share library books. But eukaryotes have a different and more complex way of passing on their genes, and they pass them on only to their offspring, never to strangers.

In eukaryotes, the genetic material is locked inside the protected vault of the nucleus. That material is released only under the most stringent conditions, using rules less promiscuous and more orderly than those of prokaryotes, and these rules affect how eukaryotic cells evolve. When eukaryotes produce germ cells — eggs and sperm, the cells from which their offspring will be formed — they don't *just* copy their DNA. They stir it around first. They swap some of their genetic material with another individual of their species so that the offspring of the two parents gets a random selection of genes, one half from one parent and the other half from the other parent. Both the genetic and the physical mechanisms involved in this elaborate dance are exquisitely complex. But the result was to add a new twist to evolution. Slight but random genetic variations were guaranteed every generation, because even if most of the genes were the same (after all, both parents are from the same species), a tiny number were always slightly different. With more variation to select from, evolution had more options. That's why evolution seems to have sped up in the past billion years. The boring billion years of the Proterozoic eon prepared the way for a much more exciting time — the Phanerozoic eon, the era of big life.

CHAPTER 6

Big Life and the Biosphere

Animals may be evolution's icing, but bacteria are the cake.

— ANDREW KNOLL, *LIFE ON A YOUNG PLANET*

Big Life

Little life ruled the biosphere for three and a half billion years and still rules much of it. It took three billion years to get from Luca to the first specimens of big life—the first multicellular animals, or metazoans. That tells us that evolving multicellular organisms was much trickier than evolving prokaryotes. And that suggests that if there is a lot of life in the universe, metazoans must be rare. Metazoans represent a new level and type of complexity among living organisms.

Many molecular mechanisms had to be in place before you could think of building multicellular organisms. You needed reliable ways of binding millions of cells together in precise structures; you needed new communication channels between cells, new ways of training cells for particular roles, new ways of managing and sharing information and energy among billions of cells. And you needed machinery that could build wings, eyes, claws, hearts, feelers, tentacles, flippers, shells, skeletons, and—because large organisms took in, processed, and responded to much more information—brains. That's a lot of new infrastructure.

It took time for this machinery to evolve, so to build metazoans, planet Earth needed one more Goldilocks condition: stability. Life-friendly conditions are not enough. You also need those conditions to persist for a long time so that life can keep evolving and experimenting. A stable sun helps here, and our sun fit the bill nicely. By stellar standards, it's a solid citizen, unlikely to do anything too unpredictable. Erratic orbits mean wild climatic gyrations, so stable planetary orbits help. Our Earth ticks this box, too. Our unusually large moon helped stabilize Earth's orbit and tilt. And, as we have seen, plate tectonics, erosion, and then life itself provided thermostats that stopped temperatures from wobbling too much at the Earth's surface.

So much could have gone wrong. A supernova in a neighboring star system could have blown up. Or we could have collided fatally with another planet. Somehow or other, our Earth avoided these dangers and remained life-friendly for more than three billion years. That was enough time for big life to evolve. And big life really is big. We are to bacteria what Dubai's 830-meter-tall Burj Khalifa is to an ant crawling past the doorman's shoes.

Once it appeared, big life would transform the biosphere as much as little life did, but in new ways. Metazoans colonized the continents and transformed them. Large plants ground rocks into soils, speeded up weathering, and turned the dusty, rocky surfaces of the early Earth, with their stromatolite-fringed shorelines, into the lush and exotic gardens, forests, and savannas of the past half billion years. As they pumped oxygen into the air, green plants on land transformed the atmosphere. Starting about four hundred million years ago, Earth got used to a new atmospheric norm of high oxygen levels (above 15 percent of the atmosphere, as opposed to the previous norm of under 5 percent) and low carbon dioxide levels (a few hundred parts per million, as opposed to a few thousand parts per million).

Animals roamed the new niches created by large plants, and fungi and bacteria cleaned up, broke down, and recycled the remains of the dead. Metazoans transformed the oceans, too, filling them with strange new creatures, from shrimps to seahorses, from octopi to blue whales.

The Molecular Gadgets That Made Big Life Possible

In the past billion years, the most important cellular innovations were not *in* the cells (prokaryotes had done most of the work here), but in the changing architectural relationships *between* cells. The earliest multicellular organisms were made from cells that bonded weakly, like the billions of cells in a stromatolite. They were really herds rather than organisms. Indeed, many bacteria show herdlike behavior, which implies some sort of rudimentary communication system. In practice, this means that the computational networks within each cell are connected into a computational system made up of many distinct cells.

Some early metazoans may have been part-time metazoans, like modern slime molds. *Dictyostelium* is an amoeba. Most of the time, its cells live independent lives. But when food is short, thousands of cells will gather together to form a slug, a larger entity that can move in search of food. And the slug can do things the individuals cannot do, such as moving large distances toward heat and light. As the slug travels, individual cells may change and take on different roles, some as spores, some as part of the stalk or foot. *Dictyostelium* tells us several important things. First, multicellularity has evolved many times and is still evolving today in some groups of organisms. Second, multicellularity, like life, has a gray border zone of organisms that are hard to classify.[1] Third, multicellularity multiplies the computational power of individual cells, increasing their ability to manage information about their environments.

In fully multicellular organisms, each cell is so specialized and so interdependent that it cannot survive alone. Genuine multicellularity is really an extreme form of symbiosis. But collaboration is made easier by the fact that most cells in metazoans are genetically identical. They are family. So each cell works to support the whole organism, sometimes by sacrificing its own life for the good of the rest. Indeed, cells, like kamikaze pilots, often self-destruct if they are no longer working well or no longer needed, a process known to biologists as *apoptosis*. Today, as many as fifty billion cells in your body will commit suicide by apoptosis.

Exchanging information is as crucial to multicellular organisms as it is to modern societies. Much intercellular communication is carried out by the cellular equivalent of a postal service; courier molecules squeeze through the membranes of individual cells and cycle between cells carrying nutrition, warnings, information, and orders. How much of the metazoan genome is devoted to collaboration became clear when the first metazoan genome was sequenced, in 1998. The organism was a worm, *Caenorhabditis elegans,* which has a nervous system with exactly 302 neurons. It turns out that about 90 percent of its 18,891 genes are not present in single-celled prokaryotes, because the job of these genes is to help cells work together.[2]

The cells of a large organism work well together because they share the same genes, but they play different roles because different genes are activated in different cells. As the single cell of a fertilized egg divides and multiplies, new cells activate different parts of their shared genome, depending on where they find themselves in the evolving embryo. Various genes determine what structures they have and what roles they will play within the organism. Managing this remarkable process of development is a small group of genes known as tool-kit genes, such as the two hundred or so *Hox* genes.[3] Tool-kit genes are like building-site managers. While ordinary genes do standard construction jobs by forming this protein or activating that enzyme, the tool-kit genes decide when and where particular molecular workers will go, using architectural plans stored in the cell's DNA. They might say, "Okay, over there, you need to start sprouting a leg," or "No, you're a bone cell, not a neuron." This is how muscle cells are created, and nerve cells and skin and bone cells and all of the two-hundred-odd different cell types that make up a human body.

The tool-kit genes are remarkably similar in different species, which suggests they are part of the earliest gadgetry of big life. It is not the tool-kit genes themselves that differentiate cockroaches from cockatoos but variations in how they go about the work of activating genes. In this way, what is a leg in one species may turn up as a wing in another species, and what began looking like a tadpole may end up as a blue whale. If tool-kit genes activate genes in the wrong order, you can get monsters, such as fruit flies with legs growing from their foreheads. The different architectural plans used by tool-kit genes help explain the remarkable variety of metazoan organisms today.

Big Life Takes Off: The Ediacaran and Cambrian Periods

Metazoans didn't flourish before about one billion years ago. The first were probably photosynthesizing algae that formed kelplike structures. But at the end of the Proterozoic eon, about six hundred million years ago, big life took off as millions of metazoan species began to explore the many new niches and lifeways opened up by multicellularity.

The rise of big life was driven by extreme climatic swings late in the Proterozoic eon. There were probably two more snowball-Earth episodes, driven by rising oxygen levels. So significant was the cold spell that began about seven hundred million years ago that in 1990, geologists added a new period to the geological timeline: the Cryogenian period. This started about

720 million years ago and lasted 85 million years. Kilometer-deep glaciers spread over the land and oceans; surface temperatures may have fallen to negative fifty degrees Celsius and photosynthesis would have largely shut down. Once again, the fate of all living organisms hung in the balance.

Why did Earth freeze? Algae spreading on land may have drawn down lots of carbon dioxide,[4] but the changing configuration of the continents may have played a role, too. Since the early Proterozoic eon, tectonic plates have periodically assembled into huge supercontinents. The supercontinent of Columbia reached its largest size about 1.8 billion years ago.[5] A billion years ago, most continents were joined together in another supercontinent known today as Rodinia. The breakup of Rodinia created a more complex global geography and sped up weathering, which would have drawn down a lot more carbon dioxide. There may have been even more violent processes at work. One possibility is a sudden shift in Earth's axis of rotation, which would have altered the position of all continents relative to the poles. Such events are known as *true polar wander events,* and they have happened at least thirty times in the past three billion years. A geological hiccup on this scale could have been caused by the sudden movement of huge masses of molten magma inside the Earth, or perhaps by an asteroid impact.[6]

Whatever the cause, these violent changes would have forced the evolutionary pace of life. Beneath the ice, surviving organisms once again huddled around cracks in the Earth's crust that leaked hot magma. In these biological refugee camps, evolution could explore odd pathways, because new genes can spread fast in small, isolated populations. Indeed, these strange worlds may have witnessed some of the earliest experiments in multicellularity.

The extreme cold ended about 635 million years ago, and it ended suddenly. Greenhouse gases from volcanoes accumulated beneath the ice and then were released explosively into the atmosphere. Carbon dioxide levels soared, while oxygen levels plummeted to well below today's levels. Temperatures rose, the ice melted, and the biosphere was transformed. Now the biological novelties that made multicellular life possible, many of them hatched in the cold, dark world of the Cryogenian period, were unleashed on a warming world.

We get the first good evidence for large numbers of multicellular organisms early in the Ediacaran period, which lasted from around 635 million years ago to around 540 million years ago. For the first time, we see the three familiar groups of large organisms: plants, which depend on photosynthesis so they can usually sit still and suck up sunlight; fungi, which scavenge decomposing organic material; and animals, which have to be alert and mobile because they survive by hunting down and eating other organisms. With the emergence of huge numbers of organisms that got their energy by consuming other organisms, the biosphere became more complex, more diverse, and more hierarchical as energy from sunlight was passed through different trophic levels, from plants to animals and fungi. Animals, such as us humans, get our energy secondhand. We use energy that was first captured by plants, and by the time it reaches us, a lot has leaked away. Ecologists talk of a food chain, a sort of queue of energy consumers with plants at the front, followed by herbivores (or creatures that consume plants), then by carnivores, which can consume herbivores, then by fungi, which bring up the rear by feasting on the dead. The whole process delights entropy, which exacts a garbage tax at every step. Approximately 90 percent of the energy captured by photosynthesis is lost at each trophic level, so much less energy is available for the later links on the food chain. That's why you find fewer animals than plants on Earth, and fewer carnivores than herbivores. But the fungi do well either way, as they recycle corpses.

The first multicellular organisms were probably plants, because they had chloroplasts inside their cells, so they could do photosynthesis. Multicelled animals evolved later, because they lived higher up the food chain, where energy was scarcer, and they needed more energy to hunt down their food. The earliest evidence for multicellular *animals* comes from the oceans of the Ediacaran period.

The Ediacaran period is named after the Ediacaran Hills in South Australia, where the first fossils from this period were discovered, in the 1940s. Paleontologists have found at least a hundred different Ediacaran genera. When initially discovered, they came as a surprise because for more than a century, biologists had assumed that the first large organisms appeared in the Cambrian period, between 540 and 490 million years ago. Biologists missed the Ediacaran creatures because most were soft-bodied, like modern sponges or jellyfish or sea anemones, so they did not fossilize well. Today, we know them mainly from the tracks and tunnels they left behind as they trudged, slithered, and burrowed through the mud of Ediacaran seas. The first cnidarians and ctenophores (think jellyfish, though that's not all these groups include) probably cruised the Ediacaran oceans. They are important to us because they are the first large organisms with nerve cells, though these were not yet concentrated into a single nervous system or brain but distributed throughout their bodies, like the nervous systems of modern invertebrates.

Biologists describe the sudden appearance of a lot of new species as an *adaptive radiation*. It's an important idea. A new biological gadget had been found—multicellularity—and now its possibilities were being explored by many different evolutionary lineages. As is the way with prototypes (think of the first internal-combustion horseless carriages), most new models didn't survive. Few Ediacaran species have obvious descendants today, and most vanished around 550 million years ago. In case you are tempted to see this as a sign of evolutionary failure, it is worth remembering that we humans have been around for just two hundred thousand years.

The Ediacaran was a sort of test run for multicellularity. The Cambrian period, which followed it, marks the start of what biologists call the Phanerozoic eon, the eon of big life, which has lasted from then until the present day. In the Cambrian period, there was a second adaptive radiation of metazoans.

Cambrian fossils were first identified in the mid-nineteenth century by an English scientist, Adam Sedgwick. At the time, Cambrian strata were the oldest to show any evidence of life. They contained many large fossils, mainly trilobites. Trilobites were arthropods, modular organisms with external skeletons, like modern insects and crustacea. Cambrian fossils were well preserved because so many had skeletons and shells. To nineteenth-century paleontologists, life seemed to pop up, fully formed, which delighted those who believed in a creator god. Now we know that life had already been around for three and a half billion years; it was just hard to see the evidence. What the Cambrian era marks is not the beginning of life but an exuberant adaptive radiation of multicellular life-forms.

Cambrian-period designs would prove more successful than those of the Ediacaran, rather as if some major glitches had been ironed out. One of the most successful design tricks of this period was modularity. You join together body modules that are pretty similar to form, say, a wormlike creature. Then the tool-kit genes start modifying each module so that some sprout legs or wings while others turn into a head with a mouth, or antennae, or perhaps a brain. Even you and I are modular, though our modules are now so specialized that it's hard to see the similarities between them.

So successful were Cambrian designs that all the major groups (or *phyla*) of large organisms existing today made their first appearance in the Cambrian period. Most showed up in the astonishing ten-million-year interval starting 530 million years ago. In this period (a split second to a paleontologist) was concentrated perhaps the most rapid stretch of biological innovation in the past six hundred million years.[7]

Cambrian species included the first Chordata, or vertebrates. This is the large phylum of animals that we belong to. Vertebrates are like tubes. They each have a spinal cord, a front (with a mouth), and a back (with an anus). They also have a rudimentary nervous system. The earliest vertebrates didn't yet have the concentrated ball of neurons that we call a brain, but they did have nervous systems with hundreds or thousands of networked nerve cells that could process lots of information fed in from sensor cells, then pass on decisions to other organs that could take appropriate action. Metazoans with even simple nervous systems could read and respond to much more information than single-celled organisms. So the Cambrian also marks the beginning of an era in which information processing became more elaborate and more important. The modern marine invertebrates called lancelets, which have nervous systems but no real brains, may bear some resemblance to our earliest vertebrate ancestors.

Unstable climates may explain the remarkable pace of evolution in the Cambrian period. Oxygen levels began to rise again, supplying some of the energy needed to form multicellular organisms. But carbon dioxide levels rose much faster, reaching levels much higher than today. This was a warm, humid greenhouse world. Whatever the exact changes, violent climatic and geological swings would have increased the evolutionary pace, driving many species to extinction and forcing the evolution of many new types of large organisms.

Evolutionary Ups and Downs: Mass Extinctions and Evolution's Roller Coaster

Like explorers crossing a mountain barrier into a new land, the invention of multicellularity opened up new possibilities for life. Metazoans explored those possibilities in multiple adaptive radiations. New life-forms transformed Earth's crust as skeletons and shells made of calcium carbonate accumulated to form thick layers of chalk (think the white cliffs of Dover). Large plants and animals moved to the land, accelerating weathering and erosion and crumbling rocks to create Earth's first true soils. Eventually, the chlorophyll in plant cells turned much of the land green.

These changes did not take the smooth, stately forms that Darwin and his generation expected of evolution. Instead, the history of big life was an unpredictable and dangerous roller-coaster ride. Asteroid impacts, sudden shifts in Earth's innards, changes in the planet's atmosphere, and massive volcanic eruptions sent evolution careering down new and unexpected pathways. Evolution was "punctuated," as Niles Eldredge and Stephen Jay Gould argued in a famous article published in 1972.[8] Like the cliché about the life of a soldier, evolution in the Phanerozoic meant long periods of boredom punctuated by moments of terror and life-threatening violence. The violence is most apparent in periods of mass extinctions.

Once again we see chance and necessity at work. At any given time, many different mixes of species were theoretically possible. Chance events determined which of these species would actually exist. During mass extinctions, whole groups of species vanished suddenly and apparently randomly. Like human wars, mass extinctions took a horrifying toll. They were particularly rough on specialized species because extreme specialists, like modern koalas, had little room to maneuver in periods of rapid change. Mass extinctions were also hard on the largest organisms, which need more food and reproduce too slowly to keep up with rapid changes. Mass-extinction events reshuffled the genetic deck of cards, created new evolutionary spaces for survivors, and set up new evolutionary experiments. They were always followed by adaptive radiations, periods of rapid experimentation during which new biological products were launched in the mass market of a changing biosphere. Many of the more exotic experiments would quickly vanish, leaving behind only the most successful.

The first mass extinctions happened back in the Archean eon. The great oxygenation event, 2.5 billion years ago, surely killed off many bacterial organisms for which oxygen was toxic. Indeed, this may have been the greatest mass extinction of all. Many groups of species also perished during the snowball-Earth episodes late in the Proterozoic eon, and we know that many disappeared at the end of the Ediacaran period. Since then, we know of at least five mass-extinction events during which more than half of all existing types of species disappeared.

The Cambrian explosion ended in a series of extinction events starting about 485 million years ago. Many species of trilobites walked the plank. So did many of the stranger Cambrian species, whose fossils have been found in the Burgess Shale in Canada and in the Chengjiang region in China.[9] The Ordovician period also ended in a mass-extinction event 450 million years ago, when 60 percent of all genera may have vanished.

The greatest of all mass extinctions came at the end of the Permian period, 248 million years ago. This time, more than 80 percent of all genera vanished, including the last of the trilobites. The precise causes of this mass extinction remain uncertain. It might have been due to rising magmas that broke through the crust in massive volcanic eruptions that sent enough ash into the air to block photosynthesis. We find modern evidence of this in a large volcanic region of Siberia known as the Siberian Traps. The eruptions pumped huge amounts of carbon dioxide into the atmosphere, so when the dust settled, carbon dioxide levels spiked, oxygen levels fell, and the oceans warmed. When Earth burped, the biosphere shuddered. By some estimates, oceans may have been as warm as thirty-eight degrees Celsius, hot enough to kill most marine organisms and stop nearly all photosynthesis in the seas. Warmer oceans could hold less oxygen and support less life, and deep beneath their surface, thawing balls of frozen methane known as clathrates may have released huge bubbles of methane. This was a *greenhouse mass extinction;* it killed by heating rather than freezing.[10] In an extreme greenhouse world, large organisms survived only in the cooler polar environments in the far north and south of the vast supercontinent of Pangaea.

Greening the Land and Oxygenating the Atmosphere

Beneath the violent changes of the early Phanerozoic, a new biosphere was building. The spread of plants, fungi, and animals onto land transformed the Earth's surface. Particularly important was the spread of photosynthesizing plants onto land, because they consumed huge amounts of carbon dioxide and released huge amounts of oxygen. That reset the biosphere's thermostats, creating a new climatic regime with higher oxygen levels and lower carbon dioxide levels than ever before. In its essential features, that regime has lasted until today.

Colonizing the land was extremely difficult, a bit like colonizing a new planet. Life had evolved and flourished in water for three billion years. Every cell had evolved in a bath of salty water. Organisms floated in water, extracted from it the gases and chemicals they needed, and fished in it for their food. Away from water, they needed support systems as elaborate as any space suit. They needed tough skins that could hold in water and prevent their bodies from drying out. But those skins also had to be permeable enough to let in carbon dioxide or oxygen. There was a tricky balance here. Leaves handle these opposite demands by way of tiny pores called *stomata* that allow carbon dioxide in and let water leak out. The size and number of stomata change depending on the surrounding temperature, humidity, and carbon dioxide levels.

How could organisms reproduce out of water? How could they protect eggs or infants from the terrible fate of desiccation? Water also provided buoyancy, and there wasn't much buoyancy on land. For tiny insects like fleas, this didn't matter. They were too light to worry about gravity, which is why a flea can happily jump off a cliff. But for large organisms, gravity *was* a problem. They needed bracing from girders of bone or wood if they were to stand up. Once standing, they needed elaborate plumbing through which liquids could be circulated against gravity to every cell in their bodies. Plants circulated liquids through roots and internal channels, exploiting water's ability to clamber upward through narrow passages using capillary action. Animals developed special pumps (aka hearts) to circulate liquids and nutrients and remove toxins.

Serious colonization of the land by metazoans began only after the late Ordovician extinction, 450 million years ago. That's when, for the first time, a few intrepid groups of plants and animals tiptoed out of the oceans and onto the land, encouraged, perhaps, by the energy boost from increasing levels of atmospheric oxygen.

The first vascular plants, with tissues that could circulate liquids and nutrients, showed up on land about 430 million years ago. Fungi and animals soon followed them. Simple, scorpionlike arthropods may have flourished on land as early as the first vascular plants. Early amphibians certainly walked the land by 400 million years ago, the date of amphibianlike fossil footprints found in Ireland and Poland. Amphibians evolved from fish that could breathe out of water and walk in the shallows of drying lakes and rivers, like modern lungfish. But all amphibians have to stay near water, where they lay their eggs. The first amphibians were the first large, land-based vertebrates. Some were as large as you and I.

Land-dwelling plants had a particularly large impact on the atmosphere, as they inhaled carbon dioxide and exhaled oxygen. Levels of atmospheric oxygen rose fast after the Ordovician period, increasing from about 5 to 10 percent of the atmosphere to levels much higher than they are today, perhaps to 35 percent, before stabilizing. Since about 370 million years ago, oxygen levels have mostly remained between 17 percent and 30 percent of the atmosphere.[11] We know this because over this entire period researchers see evidence of spontaneous fires, and fires cannot ignite if oxygen levels fall much below 17 percent. Oxygen levels probably peaked during the Permian period (from 300 to 250 million years ago).

One indicator of rising oxygen levels was the appearance of coral reefs, which need huge amounts of oxygen. The first large coral reefs appeared in the Ordovician period. Corals are really vast symbiotic colonies of tiny, genetically identical invertebrate animals. At a stretch, we might regard them as vast, sprawling animals with a hard but somewhat shapeless skeleton. Each coral hosts colonies of single-celled photosynthesizing organisms that supply it with energy. Coral reefs offered cozy lodgings for many large organisms, including trilobites, sponges, and mollusks.

Rising oxygen levels fueled a second wave of metazoan colonizers of the land during the Devonian period, which started about 370 million years ago. The first plants with woody skeletons that allowed them to stand up against gravity appeared about 375 million years ago, and the first forests appeared soon after. They fixed huge amounts of carbon through photosynthesis, so as the Earth turned green, carbon dioxide levels fell to perhaps a tenth of earlier levels.[12] The impact of the first forests was particularly significant because as yet, there were no organisms that could break down the lignin in wood. That's why forests from the Carboniferous period (from 360 to 300 million years ago) were mostly buried beneath the soil, along with the carbon they had drawn down from the atmosphere. Over time, they fossilized to form the coal seams that later powered the industrial revolution. About 90 percent of today's coal deposits were buried during the period of high oxygen levels, from around 330 to 260 million years ago. With plenty of oxygen, forest fires were easily ignited by lightning strikes. So the Carboniferous and early Permian world, though chilly, probably had the acrid smell of forest fires, a smell no one will detect on other planets in our solar system because they lack the high oxygen levels and the woody fuel sources needed for the propagation of fire.

Carboniferous forests may have doubled rates of photosynthesis, and that effectively doubled the biosphere's total energy budget, allowing the production of many more organisms.[13] Plants tweaked Earth's geological thermostat, because they sped up the weathering of rocks by grinding and dissolving them into soils that could carry buried carbon more easily into the oceans; from there, some carbon was subducted deep into the mantle. Buried carbon could no longer react with oxygen to form carbon dioxide, so oxygen levels rose. This is why the amount of free oxygen depends roughly on the amount of carbon subducted into the mantle, so levels of atmospheric oxygen and carbon dioxide tend to move in opposite directions. Rising oxygen levels also allowed new chemical reactions in the crust, creating many of the four thousand different types of minerals found on Earth today.[14]

Between 450 and 300 million years ago, from the end of the Ordovician period to the beginning of the Permian, forests and land-based metazoans transformed Earth's surface, turning the continents green and resetting the biosphere's thermostats to create the Late Phanerozoic atmospheric regime of high oxygen levels and low carbon dioxide levels.

Long Trends: Larger Bodies and Bigger Brains

Like the history of complexity in general, the history of big life was shaped by chance and necessity. Mass extinctions illustrate the dramatic role played by chance. Without them, today's biosphere would look very different. But evolution was never a matter of chance alone. Some changes were more likely than others. So, though serendipity shaped the history of big life, there were also large trends that persisted despite the turmoil caused by asteroid impacts, volcanic eruptions, and mass extinctions. The long trends are as important to us as the sudden catastrophes.

One long trend was toward bigness. That's the trend that gave us metazoans in the first place. It also encouraged the evolution of larger and larger metazoans, because being a giant often made good evolutionary sense. After all, larger organisms have fewer predators. Try getting your teeth into a blue whale! Large organisms also need less food for each unit of body weight, and it's usually easier for them to avoid the catastrophe of desiccation.[15] Besides, the high-oxygen atmospheric regime that emerged early in the Phanerozoic eon provided the extra energy needed to power megametazoans. Indeed, it seems likely that very large organisms flourished best when oxygen levels were highest, which usually meant during periods of low carbon dioxide levels and cooler climates. This was true in the oceans as well as on land, because cold water can hold more oxygen than warm water.

As oxygen levels rose, many different evolutionary lines experimented with larger bodies. During the Carboniferous and Permian periods, we begin to see mega-insects and mega-vertebrates. This was when you might have seen dragonflies with fifty-centimeter wingspans, or ninety-centimeter-long scorpion-like creatures weighing twenty kilograms. The first reptiles appeared in the Carboniferous period, which started about 320 million years ago. They were part of a new group of animals, the amniotes, which include reptiles, birds, and mammals. Unlike amphibians, amniotes could reproduce away from the water because their young developed in protected eggs, pouches, or wombs. Reptiles would eventually include some of the largest animals that have ever strolled, waddled, lolled, or galloped across the land.

The mass extinction at the end of the Permian period was followed by a new adaptive radiation during the Triassic period (from 250 to 200 million years ago). This is when we see the first large dinosaurs. (Not all dinosaurs are large!) In the later Triassic period, though, oxygen levels began to fall once more, the world began to warm, and life got tougher for massive metazoans. The Triassic world ended abruptly two hundred million years ago in another greenhouse mass-extinction event. Those dinosaur families that survived evolved highly efficient mechanisms for breathing in an oxygen-deprived world. These mechanisms may have encouraged bipedalism (think *T. rex* and modern birds), because in bipedal reptiles, the chest is more open and breathing is not checked by motion the way it is in the waddling walk of four-legged reptiles. During the Jurassic period (from around 200 to 150 million years ago), oxygen levels rose again, until they approached the levels of today's world. And dinosaurs got bigger once more. The largest tramped over the Earth in the Late Jurassic and Cretaceous periods, between 160 and 65 million years ago. Equipped with more efficient lungs than their Triassic ancestors, they used the large amounts of energy available in an oxygen-rich atmosphere to power their huge bodies.

The first true birds evolved during the Late Jurassic period. They, too, depended on high levels of atmospheric oxygen, because as every pilot knows, flight demands a lot of energy. Archaeopteryx, one of the earliest of all birdlike creatures, left fossils that were discovered in Germany in 1861, just two years after the publication of Darwin's *The Origin of Species*. It lived around 150 million years ago and was about the size of a crow. For Darwin, its discovery offered powerful evidence for his theory of evolution by natural selection because it showed the existence of transitional species, halfway between reptiles and birds. Archaeopteryx had many birdlike features, but it also retained reptilian features such as claws, a bony tail, and teeth. Recent finds have shown that many species of birds with teeth evolved during the Cretaceous and coexisted with flying dinosaurs.

Mammals, like the other amniotes (reptiles and birds) also appeared after the Permian mass extinction. Mammals would eventually produce some giants, too, but not for almost two hundred million years. Before that, they mostly lived in modest obscurity in the shadows of a world ruled by dinosaurs. Throughout the Triassic, Jurassic, and Cretaceous (from 250 million years ago to 65 million years ago), most mammals were small, burrowing creatures, a bit like modern-day rodents.

Mammals are a class of warm-blooded animals related to the other amniotes, the reptiles and birds. But mammals differ from reptiles and birds in crucial ways. The mammal brain has a neocortex, which makes mammals superb calculators. They have fur (yes, even humans have fur, though less than most mammals), and for the most part, mammals take more care of their offspring. It was Carl Linnaeus, the founder of modern taxonomy, who first called animals in our class *mammals,* after another distinctive feature: all mammals nourish their young with milk from mammary glands. For paleontologists, the most visible distinguishing feature of mammal fossils is their teeth. Even the earliest mammal teeth have cusps so that the upper and lower teeth can mesh together, allowing them to chomp down on new types of food and grind it more efficiently than most reptiles do.

Mammals illustrate another powerful evolutionary trend, the tendency toward more elaborate information processing. This is apparent throughout the Phanerozoic but particularly among animals and most strikingly among mammals.

We have seen that all living organisms are informavores. They collect information, process it, and act on it. In the simplest organisms, including prokaryotes, the second (processing) stage is rudimentary, often amounting to little more than a sort of on/off switch, as in: "It's too hot here, so wag your flagella clockwise and move away fast." Simple pain and pleasure reflexes guide a lot of effective information processing even in simple metazoans.

But as organisms became larger and more complex, they needed more information about their environments. Natural selection equipped large organisms with a desire for more information, because good information was vital to their success. That's why, when a human solves a puzzle, the brain gets the same buzz it gets from food and sex.[16] Natural selection also gave large organisms more sensors and more *types* of sensors: for sound, pressure, acidity, light. And natural selection evolved a growing repertoire of possible responses. As the number and range of inputs and outputs increased, the processing stage became more elaborate, so more nerve cells were devoted to that task. In animals, nerves gathered in nodes, ganglions, and brains, forming networks of transistor-like switches that linked hundreds, millions, or billions of neurons that could compute in parallel. That allowed them to model important features of the external world and even to model possible futures. No brainy creature (not even you or I) is in direct contact with its environment. Instead, we all live in a rich virtual reality constructed by our brains. Our brains generate and constantly update maps of the most salient features of our bodies and our surroundings, just as climate scientists model changing environments today.[17] Those maps enable us to maintain homeostasis. They help us respond appropriately, most of the time, to the never-ending swirl of changes all around us.

Decision-making works at several different levels in brainy creatures. Some decisions need to be made quickly if there's not enough time for careful deliberation. Other decision-making mechanisms are slower and more ponderous but offer more options. The simple on/off switches of pain sensors control a lot of behavior in even the most complex metazoans. Put your hand into a flame and you will remove it before you can think about it. The emotions, dominated by the limbic system, also allow rapid decision-making by creating predispositions and preferences that drive many important decisions and get them right most of the time. Charles Darwin understood that the emotions are decision-makers that have evolved through natural selection to help organisms survive. The antelope that wants to hug lions is unlikely to pass on its genes to any offspring. The most basic emotions, those least amenable to conscious control, seem to bubble up inside us. They include fear and anger, surprise and disgust, and also, perhaps, a sense of joy. They predispose us to react in certain ways and send the chemical signals that prepare our bodies to run or focus, to attack or hug.[18] Emotions drive decision-making in all animals with large brains, and some emotions, like fear, are probably present in all vertebrates and maybe in some invertebrates, particularly the most intelligent ones such as the octopi. The preferences emotions create for particular outcomes and behaviors lie behind the human sense of meaning and ethics.

The faculty we often describe as *reason* is just one of many biological decision-makers. It adjudicates on important decisions if the brain is big enough, if there is plenty of time available, and if other systems are deadlocked and can generate no clear answers. Do I really need to waste this much energy running if that is not really a lion? Is my rival making phony threats or do I need to respond?

Sensations, emotions, and thought together create the inner, subjective world that all humans, and probably many other large-brained species, experience. The state that we describe as *consciousness* seems to be a mode of sharply focused attention summoned by the brain, as if to a court of law, when new, difficult, and important decisions have to be made. That suggests that consciousness is present to some degree in many organisms whose brains are large enough to provide the necessary working space for really complex decision-making.[19] But it is not needed for routine decisions.

Add memory to these decision-making systems, and we have the foundations for complex learning, the ability to record the results of earlier decisions and use those records to make better decisions in the future. A species of fish known as cleaner wrasse, for example, clean the teeth of fish that could easily eat them. But they have to learn which clients will not eat them and may provide a free feed from between their teeth. Memory can store the results of decisions made consciously and use them for fast, automated responses. Once you've learned how to drive a car, you don't need to think through a long to-do list when you see a red light. Your body just gets on with it. You won't even notice your foot pressing on the brake.

These elaborate decision-making and modeling systems evolved throughout the Phanerozoic eon. They evolved most spectacularly in animals, because animals have to make many more decisions than plants do. In most invertebrates, neuronal networks remained distributed throughout the body, though they were often concentrated in particular nodes or ganglia. Some invertebrates, such as the octopi, have built powerful information-processing systems from such networks; most of an octopus's neurons are in its arms. In the vertebrate line, too, many neurons reach deep into the body, where they keep in touch with sensor cells and the motor cells that carry out decisions. But as sensors multiplied and processing became more critical, increasing numbers of neurons gathered together in brains, where they became specialist information processors. Information processing was particularly important in the complex, energy-guzzling lineages of birds and mammals, though these very different types of organisms evolved different subsystems to handle big data.[20]

In mammals, the increasing importance of information

processing helps explain the evolution and growth of the cortex, the gray, outer layers of the brain. The cortex provides lots of space for calculations and a lot more calculating ability, so it allowed better problem-solving in unfamiliar situations or when other decision-making systems were deadlocked. Eventually, the brainiest mammals would evolve general information-processing and problem-solving systems that were to those of the bacterial world what the Internet is to an abacus. The evolution of enhanced problem-solving and information-processing systems would eventually lead to the information explosion unleashed by our own remarkable species.

An Asteroid Lands—A Lucky Break for Mammals

For a long time, dinosaur brawn seemed to trump mammalian brains. Then, sixty-five million years ago, everything changed in a flash.

The world of the dinosaurs vanished in just a few hours when a ten- to fifteen-kilometer-wide asteroid crashed into Earth.[21] The crash caused a major extinction event, during which about half of all genera disappeared. Geologists refer to this as the K/T event because it occurred at the border between the Cretaceous period (often abbreviated K, from the German word for "chalk," *Kreide*) and the Tertiary period, an older name for the Cenozoic era, which began sixty-five million years ago.

When the asteroid hit, it was moving at thirty kilometers a second (about one hundred thousand kilometers an hour), having taken just seconds to fly through Earth's atmosphere. We know exactly where it fell: in the Chicxulub (pronounced "Chikshulub") crater in the Yucatán Peninsula of modern Mexico. The asteroid evaporated as it punched through the crust, leaving a crater almost two hundred kilometers across. Molten rocks were hurled into the air, where they formed dust clouds that blocked sunlight for many months. Limestone evaporated, spraying carbon dioxide into the atmosphere. An area hundreds of kilometers around the impact point was stripped of life. Hundreds of kilometers beyond that zone, forests lit up in massive firestorms. At sea, a tsunami formed a wall of water that crashed down on the shores of the Gulf of Mexico and killed fish and dinosaurs hundreds of kilometers away. In the Hell Creek Formation, in Montana and Wyoming, you can find fossils of fish whose gills are full of glass from the asteroid impact.[22]

Farther away, the immediate impacts were less extreme. But within weeks, the whole biosphere had changed. Soot blocked sunlight, creating what we might describe today as a nuclear winter. Nitric acid rained from the sky, killing most of the organisms it touched. The surface of Earth would have been in total darkness for a year or two, shutting down photosynthesis, life's lifeline to the sun. When the dust thinned, and light began to return through the haze, Earth warmed fast, because the atmosphere now contained a lot more carbon dioxide and methane. A few years after the impact, the wretched survivors could start photosynthesizing and breathing again, but they did so in a hot greenhouse world.

It must have taken thousands of years for the biosphere to return to something like normalcy. Meanwhile, perhaps half of all previously existing genera of plants and animals had vanished. As is typical in such crises, large species were particularly hard hit, because they need more energy, are less numerous, and reproduce more slowly than smaller creatures. This is why the large dinosaurs perished. But modern birds are descendants of smaller dinosaurs, some of which just made it through. Smaller organisms, such as the rodentlike mammals, did slightly better, and some of these would become our ancestors.

The first evidence of the asteroid impact was picked up in rocks in Italy by geologist Walter Alvarez and his team. Geologists already knew that there were striking differences between the rocks before and after the dividing line at the end of the Cretaceous period. Fossils of plankton known as foraminifera are common in the older strata just prior to that date, but they vanish after it. What was not clear was whether the change had taken tens of thousands of years or just a year or two. In 1977, at a site near Gubbio, Italy, Alvarez's team found very high levels of the rare element iridium, dating from the very end of the Cretaceous. That was odd, because iridium is rare on Earth, although it is common in asteroids. Alvarez and his colleagues found equally high levels of iridium at many other sites in Italy, and we now know of at least a hundred similar sites around the world. It began to look as if the iridium must have been brought in by an asteroid. That suggested a catastrophic event.

At the time, most geologists were committed to the idea that all geological change was gradual, so few bought the idea. They wanted direct proof, a geological smoking gun. That turned up in 1990 when it was shown that the Chicxulub crater was of just the right size and had been created at just the right date. Since then, most geologists have accepted not only that an asteroid impact wiped out the dinosaurs, but that such catastrophic events may have occurred many times in the history of Earth. True, there is also evidence of massive volcanic eruptions around the K/T boundary, and these may have undermined the health of the biosphere, but there can be little doubt now that the fatal blow was delivered by an asteroid.

The post-Chicxulub world was the world in which our mammalian ancestors would evolve. This is the world of the Cenozoic era, the past sixty-five million years of Earth's history.

After the Asteroid: A Mammalian Adaptive Radiation

As mammals, we human beings share 90 percent of our genes, or about three billion base pairs on our DNA, with other mammals, from rats to raccoons. Somewhere among the other 10 percent of our DNA lie the genes that make us different.

Like all mammals, we are warm-blooded, which means we need more energy than most reptiles to keep our body temperature up and our brains humming. Our brains need to be powerful, because they have to generate a lot of ecological tricks to maintain these large flows of food and energy. Though the earliest mammal-like creatures were no larger than mice, they probably already nursed their young, like today's mammals, and had unusually large brains in comparison to their body size. The basic division between marsupials (mammals whose young need special protection and nourishment, often in pouches) and placentals (mammals whose young are fed within the womb through a placenta) goes back at least 170 million years.

Through the long 150 million or so years of the Jurassic and Cretaceous periods, most mammal species remained small, scuttling through the moonlit undergrowth.[23] They came in many different forms. Some were doglike, such as repenomamus, a creature large enough to eat small dinosaurs and their babies. Some swam, returning to the oceans. Some were batlike, some ate insects, some climbed trees. About 150 million years ago, the world of mammals was changed by the evolution of new types of plants to rival the conifers and ferns that had dominated the plant world so far. These were the *angiosperms*, plants with fruits and flowers, the types of plants that dominate the forests and woodlands, the parks and backyards of today. Flowering plants provided a food bonanza for those mammals with teeth designed to munch on fruit and seeds or on the many insects that also munched on flowering plants or helped pollinate them.

The asteroid impact that brought down the dinosaurs may also have killed off three-quarters of all existing mammal species. But most mammals were still small, so some sneaked through the evolutionary crisis. After the planet returned to something like normalcy, survivors of the Chicxulub asteroid found themselves in a strange new world. With the dinosaurs gone, there were new opportunities. Mammals diversified in a new evolutionary radiation, as small businesses would today if every large corporation declared bankruptcy overnight. Many mammal species went big. Within half a million years, there were cow-size herbivorous mammals and equally large mammalian carnivores. There were also primates, members of the order of tree-dwelling, fruit-eating mammals from which we are descended. Though the first primates already existed in the world of dinosaurs, they flourished only after the dinosaurs had left the scene.

There was one more crisis to be survived before mammals could take over the Earth. That was the Paleocene-Eocene thermal maximum (PETM, for lovers of acronyms), a short, sharp shock of greenhouse warming at the border between the Paleocene and Eocene epochs, about fifty-six million years ago. It was damaging enough to drive many species to extinction. The PETM is of interest today because it is the most recent period of rapid greenhouse warming in Earth's history, so it may help us understand climate change today. The parallels are eerie. The amounts of carbon dioxide released into the atmosphere during the PETM were similar to those being released today by the burning of fossil fuels, and fifty-six million years ago, the result was an increase of between five and nine degrees Celsius in average global temperatures.[24]

What drove this sudden warming? Volcanic activity was unusually intense between fifty-eight and fifty-six million years ago, and carbon dioxide from volcanoes would have increased levels of atmospheric carbon dioxide. But then something happened fast, over a period of perhaps just ten thousand years, about the time that has passed in human history since the appearance of agriculture. By the end of that period, many species of plants, animals, and sea-dwellers had vanished. The best bet at present is that polar oceans warmed to the point where methane clathrates (frozen balls of methane, which look like ice but ignite if you put a match to them) suddenly melted, releasing large amounts of methane, a greenhouse gas even more powerful than carbon dioxide. That would have heated things up very fast. If this story is correct, we need to keep a very wary eye on methane clathrates in today's polar oceans.

After a climatic spike lasting perhaps two hundred thousand years, global temperatures began a long, slow descent toward colder temperatures, with a few brief reversals. Carbon dioxide levels began to fall once more, while oxygen levels rose. Differences in temperature between the equator and the polar regions increased, and ice spread across the Arctic and Antarctic, locking up water in glaciers, so ocean levels fell.

The cooling was caused in part by changes in the orbital cycles and tilt of Earth itself. These changes are known as Milankovitch cycles, after the scientist who first described them. As Earth's orbit and tilt altered, the amount of energy reaching Earth from the sun shifted in subtle ways. Tectonic processes may also have been at work, as the Atlantic Ocean widened, and the large southern continent of Gondwanaland cracked into separate modern continents. Antarctica settled over the South Pole, providing a platform for the buildup of huge ice sheets, while the northern continents circled the polar ocean, insulating the northern polar region from warm equatorial currents. Meanwhile, the collision of the Indian plate with Asia pushed up the Himalayas, which accelerated weathering, increasing the rate at which carbon was moved from the air to the sea and into the crust.

Living organisms may also have helped chill the biosphere. In the past thirty million years, as carbon dioxide levels fell, new types of plants evolved, including the grasses that cover modern savannas and suburban lawns. They used a new form of photosynthesis—C_4 photosynthesis—that was more efficient than the C_3 photosynthesis used by trees and shrubs. Because it was more efficient, it sucked more carbon out of the atmosphere.[25]

Whatever the precise causes, the cooling trend that began about fifty million years ago has continued to the present day. About 2.6 million years ago, at the beginning of the Pleistocene epoch, the world entered the current phase of regular ice ages. The world had not been this cold for 250 million years, since Pangaea itself had split apart at the end of the Permian period. Fifty million years ago, in this post-dinosaur, post-PETM world of chilly and erratic climate changes, our primate ancestors evolved.

PART III

Us

CHAPTER 7

Humans: Threshold 6

A common language connects the members of a community into an information-sharing network with formidable collective powers.

— STEVEN PINKER, *THE LANGUAGE INSTINCT*

Humanity entire possesses a commonality which historians may hope to understand just as firmly as they can comprehend what unites any lesser group.

— WILLIAM H. McNEILL, "MYTHISTORY"

The appearance of humans in our origin story is a big deal. We arrived just a few hundred thousand years ago, but today we are beginning to transform the biosphere. In the past, whole groups of organisms, such as the cyanobacteria, have changed the biosphere, but never before has a single species wielded such power. And we're doing something else that's utterly new. Because we humans can share individual maps of our surroundings, we have built up a rich collective understanding of space and time that lies behind all our origin stories. This achievement, apparently unique to our species, means that today, one tiny part of the universe is beginning to understand itself.

Our account of human history will barely touch on the things historians usually discuss: the wars and leaders, the states and empires, or the evolution of different artistic, religious, and philosophical traditions. Instead, we will stay with the main themes of our modern origin story. We will watch the appearance of new forms of complexity, created, this time, by a new species that used information in new ways to tap into larger and larger flows of energy. We will see how humans, linked first in local communities but eventually across the world, began to transform the biosphere, slowly at first, then more rapidly, until today we have become a planet-changing species. How we humans will use our power remains unclear. But we already know that humans, and indeed the entire biosphere, stand at a moment of profound and perhaps turbulent change.[1]

How did we get here? Our modern origin story can help us get our bearings by placing human history within the much larger story of planet Earth and the universe as a whole. The view from the mountaintop can help us see what makes us different.

Primate Evolution in a Cooling World

Culturally, we humans are astonishingly diverse, and that is part of our power. Genetically, though, we are more homogenous than our closest living relatives, the chimps, gorillas, and orangutans. We just haven't been around long enough to diversify much. Besides, we are extraordinarily sociable, and we love to travel, so human genes have moved pretty freely from group to group.

We belong to the mammalian order Primates, which includes lemurs, monkeys, and great apes. And we share a lot with our primate relatives. The earliest primates almost certainly lived in trees, and young humans (I include my young self here) love climbing trees and are good at it. To climb trees, you need hands and fingers or feet and toes that can grip. If you're going to leap from branch to branch, it's a good idea to have stereoscopic vision so you can judge distances. That means having two eyes at the front of your face, with overlapping lines of sight. (Don't try jumping from branch to branch with one eye closed.) So all primates have hands and feet that can grip and flattish faces with eyes at the front.

Primates are exceptionally brainy. Their brains are unusually large relative to their bodies, and the top front layer of the brain, the neocortex, is gigantic. In most mammal species, the cortex accounts for between 10 percent and 40 percent of brain size. In primates, it accounts for more than 50 percent, and in humans for as much as 80 percent.[2] Humans are exceptional for the sheer number of their cortical neurons. They have about fifteen billion, or more than twice as many as chimpanzees (with about six billion).[3] Whales and elephants, the next in line after humans on the most-cortical-neurons list, have about ten billion cortical neurons, but they have smaller brains than chimps relative to body size. Large brains mean that primates are wizards at acquiring, storing, and using information about their surroundings.

Why are primate brains so big? This may seem (pardon the pun) a no-brainer. Aren't brains obviously a good thing? Not necessarily, because they guzzle energy. They need up to twenty times as much energy as the equivalent amount of muscle tissue. In human bodies, the brain uses 16 percent of available energy, though it accounts for just 2 percent of the body's mass. That's why, given the choice between brawn and brain, evolution has generally gone for more brawn and less brain. And that's why there are so few very brainy species. Some species are so disdainful of brains that they treat them as an expendable luxury. There are species of sea slugs that have mini-brains when they are young. They use them as they voyage through the seas looking for a perch from which they can sieve food. But once they've found their perch, they no longer need such an expensive piece of equipment so . . . they eat their brains. (Some have joked, cruelly, that this is a bit like tenured academics.[4])

However, primate brains do seem to pay their way. They are needed to manage those dexterous hands and feet. And in a very visual species, they are needed to process images (is that a ripe plum three trees away?), because images gobble up processing power in brains just as they do in computers. Even more important, primates are sociable, because living in groups provides protection and support. The pressure to live in large groups increased in open and exposed terrain such as the spreading grasslands and woodlands of a cooling, post-PETM world. To live successfully with other members of your species, you have to keep track of the constantly changing relationships among family, friends, and enemies. Who's up and who's down? Who's friendly and who's not? Who owes me favors, and who am I in debt to? These are computational tasks whose complexity increases exponentially as groups get larger. If there are just three others, you can probably cope. If there are fifty or a hundred, the calculations are a lot trickier.

To live in groups, you also need some insight into the brains of others. Intuiting the thoughts and feelings of others may have been an important step toward consciousness, the enhanced awareness of what is happening in our own minds.[5] Close observation of primate societies shows that if you get these social calculations wrong, you'll probably eat less well, be less well protected, get beaten up more often, and lower your chances of being healthy and having healthy children.[6] So sociability, cooperation, and brainpower seem to have evolved together in the history of primates. Indeed, there seems to be a rough correlation between the size of primate groups and the size of their brains. Apparently, many primate lineages were willing to pay one more entropy tax, the brain tax, if it allowed them to live in larger groups.

The first primates probably evolved before the dinosaurs were wiped out, but the earliest surviving primate fossils date from several million years after the Chicxulub landing. We belong to the group of large tailless primates known as apes. Apes evolved about thirty million years ago and flourished and diversified in Africa and Eurasia twenty million years ago. The great apes (or *hominids*) include, today, the orangutans, gorillas, and chimpanzees, as well as humans. Their ancestors evolved in a post-PETM world of falling carbon dioxide levels and chillier and less predictable climates. Climatic instability pressed hard on the evolutionary accelerator, forcing many different species to adapt fast and often. From about ten million years ago, climates became drier and chillier over much of the range of great apes, and the ape lineage was culled, perhaps quite severely, as their forest homes were replaced by grasslands. Our ancestors were survivors of this evolutionary forced march.

Before the 1970s, most paleontologists were convinced from the fossil evidence that humans had diverged from other apes at least twenty million years ago. But in 1968, two geneticists, Vincent Sarich and Allan Wilson, showed that we could estimate when two species diverged by comparing the DNA of species that are alive today. This is because large stretches of DNA, particularly those parts that do not code for genes, change randomly and at a relatively consistent pace. Genetic comparisons using these insights showed that humans, chimps, and gorillas shared a common ancestor until about eight million years ago, at which point the ancestors of modern gorillas decided to go their merry way. Humans and chimps shared a common ancestor up to about six or seven million years ago. In other words, somewhere in Africa six or seven million years ago, there existed a creature from which modern humans and chimpanzees are both descended. We do not yet have fossil remains of this creature, but modern genetics tells us it was really there.

Modern chimps and humans still share well over 96 percent of their genomes. But with three billion base pairs in each genome, that means that about thirty-five million genetic letters, or base pairs, are different. Lurking among these divergent genetic letters are the clues that can tell us why humans and the chimps have had such radically different histories, particularly in recent millennia. Why are our closest relatives now reduced to remnant populations of a few hundred thousand while there are now more than seven billion humans, and we dominate the biosphere?

Early Hominin History: When Did the First Humans Appear?

All species on the human side of the evolutionary divide between humans and chimps are known as *hominins*. In the past fifty years, paleontologists have found fossil remains (sometimes just a finger bone or a few teeth) from perhaps thirty or more species of hominins. I say *perhaps* because deciding what is a distinct species depends on which paleontologist you talk to. Some are splitters; they see many different species of hominins. Others are lumpers; they see fewer species but a lot of variation within each species. Today, we are the only surviving hominin species. That is unusual, because until as recently as twenty or thirty thousand years ago, several different species of hominins cruised the savannas of Africa and Eurasia at the same time. The recent disappearance of other hominin species as we humans took up more and more land and resources is a sign of how dangerous we are.

In the past fifty years, paleontologists have acquired a lot of new forensic toys and tricks that have helped them fill in more details of hominin history. Fossilized teeth are particularly informative. That's good, because teeth are often the only remains we find. Just as your dentist can tell if you've been eating popcorn, chocolate, and ice cream, so, too, a good paleontological "dentist" can tell whether our ancestors were eating meat or plants. The shape of a tooth can tell us whether it was used to cut or grind its owner's food, and that is very informative. Nuts require grinding teeth, such as molars, while meat requires cutting teeth, such as canines.

Chemical signals found in bones and teeth can also tell us a lot about diets and lifeways. For example, C_4 photosynthesizers, such as grasses and sedges, absorb more of the slightly heavier carbon isotope, carbon-13, than they do of the more common carbon-12. Analysis of the teeth of *Australopithecus africanus* from about 2.5 million years ago shows higher than expected carbon-13 ratios, and, as they surely were not eating grasses (no apes can eat grasses), this suggests that they were eating the meat of animals that *were* eating grasses. And meat-eating implies they were either scavenging or hunting, and perhaps using stone tools.

Chemical analysis of strontium isotopes in bones can even tell us how widely individuals roamed.[7] Studies of the bones of a group of early hominins known as australopithecines have shown that females traveled more than males, which suggests that females joined groups of males rather than the other way around. In other words, their communities were patrilocal, like those of modern chimps, and that tells us a lot about their social world. These are powerful sleuthing tools. But unfortunately, they often yield more questions than answers, reminding us how complex the story of human evolution really is.

The fossil record of hominins is much richer than it used to be. In 1900, anthropologists had fossil remains from only two ancient types of humans: Neanderthals, the first of which was found in Germany in 1848, and *Homo erectus*, whose remains were first found in 1891 in Java by Dutch paleoanthropologist Eugène Dubois. These finds suggested that humans could have evolved in Europe or Asia. But in 1924, Raymond Dart, an Australian professor of anatomy based in South Africa, discovered the first important African hominin fossil. It was a skull sitting among a collection of other fossils, the skull of a child from the species now known as *Australopithecus africanus,* part of a large group of australopithecine species that first appeared about five million years ago. After this discovery, more and more hominin fossils began to turn up in Africa, and most paleoanthropologists now believe that our species evolved somewhere in Africa. From the 1930s, Louis and Mary Leakey began finding hominin fossils and artifacts in Africa's rift valley, where magma pushing up from the mantle has started splitting the tectonic plate on which most of Africa lies. Eventually, a new sea will appear here. Meanwhile, cracks in the African tectonic plate give fossil hunters glimpses into the remote past of our species.

In 1974, in Ethiopia, Donald Johanson discovered 40 percent of the skeleton of another australopithecine species, *Australopithecus afarensis.* The skeleton was named Lucy and dated to about 3.2 million years ago. Other australopithecine remains have been found that are almost four million years old. Since then, earlier hominin species have been found in other parts of Africa, dating to four and five million years ago (*Ardipithecus*) and even to six million years ago (*Orrorin tugenensis*), or perhaps seven (*Sahelanthropus tchadensis*), which is pretty close to the notional date when the last common ancestor of all hominins lived.

We have so few very early hominin fossils that a single new discovery could change the story radically. It is not even certain that the oldest fossils are really hominins, nor is it always clear whether fossil remains belong to distinct species or not. Should *Homo habilis* and *Homo erectus,* species with very different brain sizes, be assigned to different genera, or should *H. habilis* be regarded as late australopithecines? Our understanding of early hominin history remains sketchy, but parts of the story are getting clearer.

Even the earliest hominin species seem to have walked on two legs, at least some of the time. This is very different from chimps and gorillas, which knuckle-walk. You can tell from bones if a species regularly walks on two legs. In bipedal species, the big toe is no longer used for gripping, so it aligns more closely with the other toes, while the spine enters the skull from below, not from the back (get down on four legs and you'll understand why). Walking on two legs required rearrangements of the back, the hips, even the braincase. It also favored narrower hips, which made childbearing more difficult and dangerous and probably means that many hominins, like modern humans, gave birth to infants that were not yet capable of surviving on their own. That would have meant that their babies needed more parenting, which may have encouraged sociability and gotten hominin fathers more involved in child-rearing. There were many indirect effects of bipedalism, but we're not yet sure exactly why hominins became bipedal. Perhaps bipedalism let our ancestors walk or run farther in the grassy savanna lands that had spread around a cooling world in the past thirty million years. It also freed human hands to specialize in manipulative tasks including, eventually, the making of tools.

There are no signs that the earliest hominins were exceptionally brainy by primate standards. Their skulls contained brains much smaller than ours and more like chimpanzee brains, with a volume of about 300 to 450 cubic centimeters. Our brains, in comparison, average about 1,350 cubic centimeters. More significant than absolute size, though not easy to calculate, is the extent to which brain size deviates from the expected brain size for a given body weight within a particular group of organisms. This is the encephalization quotient (EQ). Chimps have an EQ of about 2 (compared to other mammals), and modern humans have an extraordinarily high EQ of about 5.8. The EQs of australopithecines range from 2.4 to 3.1.[8] Extreme braininess was *not* the first distinguishing feature of the hominins. Bipedalism was.

The first fossils that are currently classified within our own genus, *Homo,* belong to a species known as *Homo habilis,* which lived in Africa from about 2.5 to 1.5 million years ago. The first evidence of this species, consisting just of a jawbone and some hand bones, was found in 1960 by Mary Leakey and her son Jonathan in Olduvai Gorge in the African rift valley. The close association with stone tools persuaded the Leakeys to classify the new species as a form of *Homo*, which was a paleontologist's way of saying "I think these are really humans because they made tools."

But were they us? Is this when human history began? Today, most researchers are skeptical about a distinct *Homo* genus that includes both us and *habilis*. After all, *habilis* brains were only slightly larger than those of australopithecines, ranging from 500 to 700 cubic centimeters, with an encephalization quotient of just over 3. And their stone tools involved little more than smashing rocks and using the fragments. Given that some australopithecine species probably made stone tools and that chimps, too, can make tools (though not stone tools), it looks as if *Homo habilis* was similar enough to the australopithecines to be classified with them. Tool use does not make them human, because we now know that toolmaking is not unique to humans.

Later Hominin History: The Past Two Million Years

By two million years ago, at the beginning of the Pleistocene epoch, we find hominin species that were larger, had bigger brains, made more sophisticated stone tools, and exploited a wider range of environments. It is probably no coincidence that they appeared as climates were getting colder and drier. These species are normally classified as *Homo erectus* or *Homo ergaster*, but here I will use the label *H. erectus* for the whole group.

The large brains of *H. erectus* are striking because, as we have seen, brains are costly evolutionary machines. Indeed, the rate of increase of brain size to body weight in hominins was faster than the rates in any other group of species in evolution-ary history.[9] Perhaps sociability was the driver. The importance of social calculations shows up clearly in the human brain struc-ture, which devotes an exceptional number of neuronal path-ways to social calculations. Perhaps more neurons meant more friends, more food, better health, and a better chance of repro-ducing. Certainly, larger brains allowed hominins to live in larger groups and networks.[10] Most primates, including chimps and baboons, lived in groups of fewer than fifty individuals, and, roughly, the smaller the brain, the smaller the group. But as brain sizes increased in the past two million years, the size of hominin groups increased, too. *Homo erectus* was probably the first hominin species to live in groups that linked more than fifty individuals.

The first *H. erectus* remains were found in Java in 1891 by Eugène Dubois. He was looking in Indonesia because of a hunch he had that humans were descended not from African chimpan-zees (Darwin's bet), but from Asian orangutans. He got that wrong. But the remains he found did have brains with volumes of almost 900 cubic centimeters, much closer to the modern human average of about 1,350 cubic centimeters. And they had an EQ of 3 to 4. The fact that the remains were found in Java also showed that *H. erectus* had the technologies and skills needed to migrate from Africa through much of southern Eur-asia. But we shouldn't be too impressed by this. Many other spe-cies, such as lions, tigers, elephants, and even our close relatives the orangutans, had made similar migrations, and that's because many environments in southern Eurasia are not that different from African environments. Indeed, recent evidence suggests that species closely related to *Homo habilis* may have traveled as far as Indonesia to become the ancestors of the tiny hominins known as *Homo floriensis* (or the Hobbits), which lived on the island of Flores as recently as sixty thousand years ago.[11]

H. erectus were taller than *H. habilis,* some of them as tall as modern humans. They also made more sophisticated stone tools than *H. habilis*. These are the beautiful and carefully designed stone tools known as Acheulean axes. Better stone tools may have given *H. erectus* access to more meat, a crucial source of high-energy food to fuel their expanding brains. They may have also learned to manage, control, and use fire, which would have allowed them to tap into a huge new source of energy. The primatologist Richard Wrangham has argued that *H. erectus* used fire to cook (in other words, to predigest and detoxify) meat and other foods. This would have increased the range of foods they could eat, because many foods are indigestible or poisonous until cooked. Cooking would also have reduced the time they spent chewing and digesting their food.

Use of fire may have had other important consequences. For example, cooking reduced the digestive work required of the gut, so the gut shrank (and, yes, there is fossil evidence for this), releasing some of the metabolic energy needed to run larger brains. As yet, this interesting hypothesis remains unproven, because good evidence for systematic control of fire appears only from about eight hundred thousand years ago and becomes quite common only after about four hundred thousand years ago.[12] We also know that the stone technologies of *H. erectus* changed little over a million years, so *H. erectus* seem to have lacked the technological flair and creativity of our own species.

In the past million years, hominin evolution accelerated. About six hundred thousand years ago, new species appear in the fossil record, with brains and bodies more and more like modern humans'. Not surprisingly, they apparently lived in larger groups, too, groups that linked as many as 150 individuals, which seems to have been the upper limit among our hominin ancestors.[13]

There are complex debates about how many different species of hominins there were half a million years ago. We know there were many. But more important is the larger trend: Now hominins appear in ice-age Europe and northern Asia, environments that were very different from the African savanna and demanded new skills and technologies. So it is no surprise that their tools were more sophisticated, more varied, and more specialized than those of *H. erectus*. For the first time, hominins hafted stone points to wooden shafts. In Schöningen, Germany, archaeologists have found four-hundred-thousand-year-old wooden spears made with precision and delicacy. Some anthropologists have even detected evidence of artistic and ritual activity. Among the finds of Eugène Dubois were decorated mussel shells, dated to five hundred thousand years ago, that look suspiciously like simple forms of art.

Still... none of this was revolutionary. The really spectacular changes began only about two or three hundred thousand years ago, after the appearance of our own species, *Homo sapiens*.

What Makes Us Different? Crossing Threshold 6

Imagine a team of alien scientists who have been orbiting our planet searching for intelligent life and studying Earth's lifeforms in a longitudinal research project lasting several million years. Two hundred thousand years ago, they wouldn't have noticed anything unusual about our ancestors. In Africa and parts of Europe and Asia, they might have spotted several species of large, bipedal primates, including the species we call *Homo neanderthalensis* and *Homo heidelbergensis*. They might even have seen individuals that a modern human paleontologist would describe as *Homo sapiens*, because the oldest skull normally assigned to our species is almost two hundred thousand years old. It was found at Omo Valley in Ethiopia in the African rift valley. (In June 2017, human remains from Morocco were dated to three hundred thousand years ago, but their exact relationship to us remains uncertain.) But there was little to distinguish these early humans from many other large or medium-size primate and mammal species. They lived in small, scattered nomadic communities with a total population of, at most, a few hundred thousand individuals. Like all large animals, they gathered or hunted the food and energy they needed from their surroundings.

Today, two or three hundred thousand years later (no time at all for a paleontologist), our orbiting aliens searching for intelligent life would have seen enough changes in the behavior of this particular species to justify a few scholarly high-fives. They would have watched as humans spread around the world. Then, starting from the end of the last ice age, ten thousand years ago, they would have noticed human numbers growing fast. They would also have watched as humans began to change their environments to suit them better by burning down forests, diverting rivers, plowing the land, and building towns and cities. In the past two hundred years, human numbers grew to over seven billion, and our species began to transform the oceans, the land, and the air. Human-built roads, canals, and railways snaked across the continents, linking thousands of human-built cities with populations in the millions. Vast ships navigated the oceans, and planes ferried goods and people through the air and across the continents. Just a hundred years ago, in glowing filaments and patches, Earth started lighting up at night. The aliens' instruments would also have shown that oceans were getting more acidic, the atmosphere was warming, coral reefs were dying, and polar ice caps were shrinking. Biodiversity was declining so fast that some of the alien biologists might have wondered if this was the start of another mass extinction.

Paleontologically speaking, changes this fast are the equivalent of an explosion. Without planning it, we have become a planet-changing species. We even have the power, if we are foolish enough, to destroy much of the biosphere in just a few hours by launching some of the eighteen hundred nuclear missiles that remain on high alert today. No single species has had such power in the four-billion-year history of the biosphere.

Clearly a new threshold had been crossed. Our alien scientists would surely have been asking themselves, What *is* it about this strange species?

Historians, anthropologists, philosophers, and scholars in many other fields have wrestled long and hard with the same question. Some feel the question is too complex, too loaded, and too multidimensional to yield a scientific answer. But curiously, when we see human history as part of the larger history of the biosphere and the universe, the distinctive features of our species stand out more clearly. Today, scholars in many different fields seem to be converging on similar answers to the question of what makes us different.

When you see sudden, rapid changes like this, start looking for tiny changes that have huge consequences. Complexity theory and the related field of chaos theory are full of changes like this. Often, they are described as butterfly effects. The metaphor comes from the meteorologist Edward Lorenz, who pointed out that in weather systems, tiny events (the flapping of a butterfly's wings, perhaps?) can get amplified by positive feedback cycles, generating a cascade of changes that may unleash tornadoes thousands of miles away. So what tiny changes unleashed the tornado of human history?

Many different features make up the human package, from dexterous hands to large brains and sociability. But what makes us radically different is our collective control of information about our surroundings. We don't just gather information, like other species. We seem to *cultivate* and *domesticate* it, as farmers cultivate crops. We generate and share more and more information and use it to tap larger and larger flows of energy and resources. New information gave humans improved spears and bows and arrows that allowed them to hunt larger animals more safely. It gave them better boats that gave them access to new fisheries and new lands, and it offered new botanical knowledge that allowed them to leach the poisons from potentially edible plants such as cassava. In more modern times, new information lay behind the technologies that let us tap the energy of fossil fuels and build the electronic networks that link us into a single world system.

Information management on this scale was not the achievement of individuals. It depended on sharing, on the accumulation of millions of individual insights over many generations. Eventually, community by community, this sharing created what the Russian geologist Vladimir Vernadsky called a *noösphere*, a single global realm of mind, of culture, of shared thoughts and ideas. "There is," writes Michael Tomasello, "only one known biological mechanism that could bring about these kinds of changes in behavior and cognition in so short a time....This biological mechanism is social or cultural transmission, which works on time scales many orders of magnitude faster than those of organic evolution." This process, which Tomasello calls "cumulative cultural evolution," is unique to our species.[14]

The tiny change that allowed humans to share and accumulate so much information was linguistic. Many species have languages; birds and baboons can warn others in their group of the approach of predators. But animal languages can share only the simplest of ideas, almost all of them linked to what is immediately present, a bit like mime (imagine trying to teach biochemistry or wine-making in mime). Several researchers have tried to teach chimps to talk, and chimps can, indeed, acquire and use vocabularies of one or two hundred words; they can even link pairs of words in new patterns. But their vocabularies are small and they don't use syntax or grammar, the rules that allow us to generate a huge variety of meanings from a small number of verbal tokens. Their linguistic ability seems never to exceed that of a two- or three-year-old human, and that is not enough to create today's world.

And here's where the butterfly flapped its wings. Human language crossed a subtle linguistic threshold that allowed utterly new types of communication. Above all, human languages let us share information about abstract entities or about things or possibilities that are not immediately present and may not even exist outside of our imagination. And they let us do this fast and efficiently. With the partial exception of honeybees, whose dances can tell other bees where to find honey, we know of no animals that can transmit precise information about what is not right in front of them. No animal can swap stories about the future or the past, or warn about the lion pride ten miles to the north, or tell you about gods or demons. They may be able to think about such things, but they cannot talk about them. And that may be why it is hard to find any evidence for teaching within any other species, even among our closest relatives, the monkeys and apes.[15]

These linguistic enhancements allowed humans to share information with such precision and clarity that knowledge began to accumulate from generation to generation. Animal languages are too limited and too imprecise to allow this sort of accumulation. If any earlier species did have this ability, it would surely have left traces, including an expanding range and an increasing impact on its environment. In fact, we would see the sort of evidence we find for human history. Human language is powerful enough to act like a cultural ratchet, locking in the ideas of one generation and preserving them for the next generation, which can add to them in its turn.[16] I call this mechanism *collective learning*. Collective learning is a new driver of change, and it can drive change as powerfully as natural selection. But because it allows instantaneous exchanges of information, it works much faster.

How and why our species acquired the linguistic power needed to unleash this powerful new driver of change remains unclear. Was it, as American neuroanthropologist Terrence

Deacon has argued, a new ability to compress large amounts of information into symbols (deceptively simple words like *symbol* that carry a huge informational cargo)? Or was it the evolution of new grammar circuits in the human brain that helped us combine words according to precise rules so as to convey a great variety of different meanings, as the linguist Noam Chomsky has suggested? This is a tempting idea because, as another linguist, Steven Pinker, puts it, the really difficult trick was "to design a code that can extrude a tangled spaghetti of concepts into a linear string of words" and to do this so efficiently that the hearer could quickly re-create the spaghetti of concepts from the linear string.[17] Was human language enabled by the increased space for thinking available in an enlarged cortex, which could hold enough complex thoughts in place to form syntactically complex sentences or let an individual memorize the meanings of thousands of words?[18] Or do improved forms of language have their roots in the sociability and willingness to collaborate that is particularly well developed in our own species?[19] Or was there perhaps a synergy between all these drivers?

Whatever happened, our species seems to have been the first to cross the linguistic threshold beyond which information can accumulate within communities and across generations. Like a gold strike, collective learning unleashed a bonanza of information about plants and animals, about soils, fire, and chemicals, and about literature, art, religion, and other humans. Though some information was also lost every generation, in the long run, human stores of information accumulated, and that growing wealth of knowledge would drive human history by giving humans access to increasing flows of energy and increasing power over their surroundings. Here is how this mechanism is described by a pioneer of the study of memory, the Nobel Prize winner Eric Kandel:

Although the size and structure of the human brain have not changed since *Homo sapiens* first appeared in

East Africa...the learning capability of individual human beings and their historical memory have grown over the centuries through shared learning—that is, through the transmission of culture. Cultural evolution, a nonbiological mode of adaptation, acts in parallel with biological evolution as the means of transmitting knowledge of the past and adaptive behaviour across generations. All human accomplishments, from antiquity to modern times, are products of a shared memory accumulated over centuries.[20]

The great world historian W. H. McNeill constructed his classic world history *The Rise of the West* around the same idea: "The principal factor promoting historically significant social change is contact with strangers possessing new and unfamiliar skills."[21]

Living in the Paleolithic

Human history begins, then, with collective learning. But when did collective learning begin?

Even our alien scientists would hardly have noticed the first flickering of collective learning as they circled Earth two hundred thousand years ago. Some form of collective learning may have been at work even in *H. erectus* communities, but its consequences were not yet revolutionary. Hints of more rapid technological change begin to appear in the African archaeological record at least three hundred thousand years ago in the form of increasingly delicate stone tools, many of them hafted.[22] And it is not just *Homo sapiens* who show this creativity but also Neanderthals and the hominin species known as *Homo heidelbergensis*. Perhaps all these species were acquiring improved forms of language that brought them tantalizingly close to threshold 6.

Early evidence of ritual or symbolic or artistic activity is particularly significant because it suggests an ability to think symbolically or tell stories about imaginary beings, and that may indicate the arrival of modern forms of language.

Perhaps there was room for only one species to cross the threshold to collective learning. There is an evolutionary mechanism known as competitive exclusion that explains why two species can never share exactly the same niche. One will eventually drive out its rival if it can exploit the same niche slightly more effectively. So we can imagine several species gathering near the evolutionary threshold to collective learning, but then one broke through and began to exploit its environment so efficiently that its numbers multiplied and grew fast enough to lock out its rivals.[23] This may help explain why our closest hominin relatives, such as the Neanderthals, have perished, and our closest surviving relatives, the chimps and gorillas, are approaching extinction.

Evidence of technological and cultural change from before a hundred thousand years ago is foggy and difficult to interpret. Our own lineage began to spread within Africa starting at least two hundred thousand years ago, which may point to the advantages of collective learning.[24] But in a world of small, scattered communities, most of them little larger than extended families, change was slow, erratic, and easily reversed. Whole groups could die out suddenly, along with the technologies, stories, and traditions they had built up over many centuries. The largest catastrophe of this kind occurred about seventy thousand years ago. Genetic evidence shows that the number of humans suddenly fell to just a few tens of thousands, only enough to fill a moderate-size sports stadium. Our species came close to extinction. The catastrophe may have been triggered by a massive volcanic eruption on Mount Toba in Indonesia that pumped clouds of soot into the atmosphere, blocking photosynthesis for months or years and endangering many species of large animals. But then human numbers began to increase again; humans spread more widely, and the machinery of collective learning roared into life once more.

In the past one hundred thousand years, we get some glimpses of how our ancestors lived and find clearer evidence for collective learning. Like all large animals, our ancestors collected or hunted resources and game from their surroundings. But there was a crucial difference between those animals and early humans. While other species hunted and gathered using a repertoire of skills and information that barely changed over the generations, humans did so with increasing understanding of their environments, as they shared and accumulated information about plants, animals, seasons, and landscapes. Collective learning meant that, over the generations, human communities hunted and gathered with growing skill and efficiency.

Some sites give us intimate glimpses of how our ancestors lived. At Blombos Cave, on the Indian Ocean shores of South Africa, archaeologist Christopher Henshilwood and his colleagues have excavated sites dating from ninety thousand to sixty thousand years ago. The inhabitants of Blombos Cave ate shellfish, fish, and marine animals as well as land mammals and reptiles. They cooked in well-tended hearths.[25] They made delicate stone blades and bone points that were probably hafted to wooden handles with specially prepared glues. But they were also artists. Archaeologists have found ocher stones with geometrical scratch marks on them that look for all the world like symbols or even writing. They also made different-colored pigments and ostrich-shell beads. It is tempting to see this evidence as a sign that the Blombos communities *valued* collective learning and the preservation and transmission of information, and that surely means that they preserved and told stories that summed up their community's knowledge.

It is hard not to see similarities with modern foraging communities. If these similarities are not misleading us, we can imagine many groups like those from Blombos Cave with a great diversity of gathering and hunting techniques built up over many generations. We can imagine them migrating through familiar home territories, held together by family ties and shared languages and traditions. They surely danced and sang, too, and told origin stories, and they almost certainly had what we moderns might want to call religions.

At the Lake Mungo site in Australia, the evidence for religion is compelling. A cremation and burial from about forty thousand years ago and a scattering of other human remains are evidence of rich ritual traditions. Other evidence from the site reminds us that Paleolithic societies, like modern human societies, underwent profound upheavals, many caused by the unpredictable climate changes of the most recent ice age. There were regular periods of aridity from the moment humans first arrived in the Willandra Lakes Region, perhaps fifty thousand years ago. About forty thousand years ago, aridity increased and the lake system began to shrink.

Twenty thousand years later, at the coldest phase of the ice age, there were communities living in tundra-like environments on the steppes of modern Ukraine. At sites like Mezhirich, people built huge marquee-like tents, using skins stretched over a scaffolding of mammoth bones, and warmed them with internal hearths. They hunted mammoths and other large animals and stored meat in refrigerated pits for recovery during the long cold winters. They hunted fur-bearing animals and used needle-like objects with ornamental heads carved from bone to sew warm clothing. As many as thirty people may have lived together at Mezhirich during the long ice-age winters. There are similar sites near Mezhirich. This suggests there were regular contacts between neighboring groups, the sort of networks through which information about new technologies, changing climates, animal movements, and other resources would have been exchanged, as well as stories. People, too, would have moved between neighboring groups.

The remains left behind by Paleolithic communities offer grainy snapshots of their societies. But each snapshot represents an entire cultural world, with stories, legends, heroes, and villains, scientific and geographical knowledge, and traditions and rituals that preserved and passed on ancient skills. This accumulation of ideas, traditions, and information was what allowed our Paleolithic ancestors to find the energy and resources they needed to survive and flourish and migrate farther and farther in a harsh, ice-age world.

Evidence from ice cores now lets us track global temperature changes with great precision across hundreds of thousands of years. During the Pleistocene epoch, which encompasses the two million years since the evolution of *Homo erectus,* there were many ice ages. They normally lasted for one hundred thousand years or more, with briefer warm periods, or interglacials, between them. The period we live in now is a warm interglacial that began ten thousand years ago, at the start of the Holocene epoch. The previous interglacial occurred about a hundred thousand years ago and may have lasted for twenty thousand years or more. After it ended, global climates got steadily colder and drier, though with many temporary reversals and local variations. The coldest period of the last ice age was from about twenty-two thousand to eighteen thousand years ago.

As climates cooled, areas that had been occupied for hundreds or thousands of years had to be abandoned. Sites in northern Europe that had been occupied starting about forty thousand years ago were abandoned for thousands of years. Even in the warmer climates of Australia's far north, people survived by the skin of their teeth.[26] Lawn Hill River in the far northwest of Queensland carved gorges through thick layers of limestone and provided local people with a good living from both the fish and marine animals of the rivers and the surrounding highlands. But during the coldest phases, people abandoned the icy highlands entirely and stayed in the protected environments of the gorges.

Settling the Biosphere: Humans Migrate Around the World

As technological and ecological knowledge accumulated, many communities moved into new environments, pulled or pushed by climate change, by conflicts with their neighbors, or, perhaps, by overpopulation. Over thousands of years, small-scale migrations would eventually take our species, kilometer by kilometer, to every continent other than Antarctica. Today, we can track these migrations by following the spread of archaeological remains around the world and by comparing the genes of different modern populations.[27]

One hundred thousand years ago, during the last interglacial, almost all humans lived in Africa, though a tiny number had left for the Middle East. At sites such as the caves of Skhul and Qafzeh in modern Israel, they may have encountered and occasionally interbred with Neanderthals. (We know this because today, most humans who live outside Africa have some Neanderthal genes.) Then, as climates cooled, our ancestors seem to have left the Middle East to the Neanderthals, whose bodies were better adapted to colder climates. They didn't return until about sixty thousand years ago. However, some humans may have traveled east into Central Asia and South Asia. One reason for thinking this is that humans reached Sahul (the ice-age continent that included Australia, Papua New Guinea, and Tasmania) between fifty thousand and sixty thousand years ago. Migrants leaving Africa sixty thousand years ago would have had to move extraordinarily fast to get there, so it seems more likely that the first Australians arrived from communities long established in Asia.[28] Settling Australia was a major event in human history. We don't know what drove the first settlers—probably population pressure or conflicts with other communities in the southern parts of what is now Indonesia. But we do know that the crossing required advanced seafaring skills and the ability to adapt fast to an entirely new suite of plants and animals. No other species made the sea crossing. (Dingoes arrived in recent millennia, almost certainly with human help.)

The earliest migrations into Siberia and northern Europe were probably short-lived exploratory probes during brief warm periods. But sites such as Mezhirich show that by twenty thousand years ago, our ancestors could cope with extremely cold environments. Some may have settled permanently in Siberia as early as forty thousand years ago. Twenty thousand years later, at the coldest phase of the last ice age, some Siberians trekked east across the land bridge of Beringia, which was crossable because so much water was locked up in polar glaciers that ocean levels were lower than today. From Beringia, humans spread into the Americas, either by going through Alaska or by traveling in small boats along the northwestern coast of North America. From there, some migrated into South America, probably reaching as far south as Tierra del Fuego within two or three thousand years. At present, the earliest firm evidence for the presence of humans in North America dates to about fifteen thousand years ago.

In the Paleolithic period, migration was probably the most common reaction to innovations or population pressure. A trickle of emigration meant that each human community could remain about the same size as our species spread around the world, and that meant that communities could preserve many of their traditional social rules. This is why we have little evidence for large Paleolithic settlements, though there is plenty of evidence that the total number of communities increased, as well as the total number of humans. The English anthropologist

Robin Dunbar has argued that 150 people represents the largest group size that human brains can normally cope with, so it may be that communities naturally split if they got any larger. Dunbar has argued that even today, most humans are embedded in intimate networks that are no larger than 150, even if they have more fleeting relationships with many other people. Modern communities are huge, but only because of the creation of special new social structures to hold them together.

Whatever the reasons, most Paleolithic communities remained small enough to organize themselves through notions of family or kinship, like most modern foraging societies. That's why it makes sense to think of Paleolithic communities as families rather than societies. And if modern foraging communities are any guide, they probably had a broad understanding of the term *family* that extended beyond the world of humans to include other species and even features of the landscape, such as mountains and rivers. Paleolithic societies were embedded in their surroundings ecologically and culturally in ways that modern urban dwellers struggle to understand.

Increasing Complexity in the Paleolithic

Though small, Paleolithic communities had the universal human knack of accumulating new ideas, insights, and knowledge, so even if it we cannot track their histories in detail, we know that they showed the same cultural and technological dynamism as later human communities did, if on a smaller scale.

Like modern foragers, our Paleolithic ancestors surely had intimate and precise knowledge of the habits and life patterns of the animals and insects they hunted and the plants they used for their food, clothing, and equipment. The looser networks through which people, stories, rituals, and information were exchanged would have linked communities over large areas. From archaeological and anthropological evidence, we can conclude that family groups lived separately for most of the time but gathered periodically in Paleolithic equivalents of the Olympic Games at sites where there was enough food to support temporary gatherings of hundreds of individuals. In the Snowy River region of Southeast Australia, for example, many groups came together when millions of bogong moths hatched, providing the food needed to support the large gatherings known today as corroborees. At these meetings, stories were swapped, rituals and gifts were exchanged, ties of solidarity were maintained in dances and ceremonies, and marriage partners (or disgruntled individuals) moved from group to group. In the south of France fifteen thousand years ago, there were similar gatherings, as human communities followed and hunted herds of horses, deer, and cattle and engaged in periodic rituals that generated beautiful rock art. The art and sculptures produced at sites such as the Lascaux Caves and the La Madeleine rock shelter in the Dordogne region, and the even older stone carvings found in many parts of Australia, are, to modern eyes, as beautiful and sophisticated as any art ever produced by humans. They help illuminate the rich intellectual and mental world of our Paleolithic ancestors.

As hunting and gathering techniques became more sophisticated, our ancestors began to shape their environments in new ways. In some parts of the world, they changed the mix of surrounding species. The first humans in Australia found many species of large animals, or megafauna. Some were as big as the rhinoceroses, elephants, and giraffes of South Africa, the one part of the world in which large numbers of megafauna survive today. In Australia, there were giant kangaroos and wombats and huge flightless birds such as *Genyornis newtoni*. Then, quite suddenly, most of the Australian megafauna disappeared, as they would eventually disappear in Siberia and the Americas.

Perhaps they disappeared because climates changed. But they had survived previous ice ages, so it is hard not to think that humans, with their increasingly sophisticated hunting methods, may have tipped them over the edge. The chronology supports this explanation. In Australia, Siberia, and North America, the megafauna vanished not long after the arrival of humans. Perhaps, like the dodo in Mauritius, the megafauna didn't fear our ancestors enough, unlike African megafauna, which had coevolved with humans and knew how dangerous we could be. In any case, megafauna, like all large animals (including the dinosaurs), are particularly vulnerable to sudden changes. There are many modern examples of megafaunal extinctions, such as the disappearance of the large New Zealand birds known as moas within a few centuries of the arrival of humans. In Siberia and the Americas, we even have direct evidence of kill sites, so we know that humans hunted megafauna such as mammoths.

Removing megafauna changed landscapes. Large herbivores can chomp their way through a lot of plants. Eliminating them increased the frequency of fires, as plant remains were left uneaten. In Australia about forty thousand years ago, the number of fires increased in many regions. A large percentage may have been started by lightning strikes. But we know that here, as in many other parts of the Paleolithic world, humans used fire systematically to fertilize the land. These technologies are known to archaeologists as fire-stick farming, after the fire sticks that indigenous Australians carried to fire the land in historical times. Systematic use of fire, not just to cook or protect yourself but to transform your environment, represents one of the first signs of the growing ecological power of our species. If you had the skills needed to manage fires safely, regular firing of the land provided many advantages. Burn an area of grassland, then wander back in a day or two, and the first thing you will find is plant and animal barbecues. Wait a few weeks and you will find new growth, because the fire has scattered ash as a fertilizer and sped up the decomposition of plant and animal remains. Grasses and other plants will sprout and can be harvested sooner. And new plants will usually attract herbivores and small reptiles, making the hunting easier and more productive. In short, fire-stick farming increases the productivity of the land.

Similar techniques were used in many parts of the world in the late Paleolithic. Though not strictly a type of farming, they *were* a way of increasing the production of usable plants and animals in a given area of land. They count, in other words, as a form of intensification. Fire-stick farming gives us a preview of the bonanza of food, resources, and energy that would be released by farming.

The Earliest Era of Human History

As people shared information, ideas, and insights, as well as jokes, gossip, and stories, over many generations and among neighboring communities, there slowly accumulated, region by region, a body of information that I am tempted to call *scientific*. Paleolithic science included knowledge about usable resources, whether hunted or gathered, whether for eating or making clothes or healing; knowledge about techniques, whether for navigation or hunting or digging for root crops; knowledge about astronomy; and social knowledge about how to approach and talk to elders or strangers and how to mark important transitions in the lives of individuals. It was valuable knowledge because it was needed for survival, so tending it and passing it on was a matter of great seriousness. Knowledge was filtered through many minds, tested for its authoritativeness, accuracy, and usefulness, and eventually incorporated in the origin stories that lay at the heart of education. And this slow increase in available information and the control that this accumulated information gave our species over the natural world and over energy flows through the biosphere would turn out to be the primary driver of change in human history. As humans spread, so did knowledge. Though knowledge was still compartmentalized, community by community, we can imagine the slow emergence, for the first time in the planet's history, of a new sphere of shared knowledge, the noösphere.

During the Paleolithic period, the noösphere expanded through Africa, Eurasia, Australasia, and then to the Americas, as human numbers increased. When human communities spread within Africa, their populations may have risen to a few tens of thousands, or even hundreds of thousands, though there were surely local fluctuations in numbers. And as we have seen, human numbers plummeted to just a few tens of thousands just seventy thousand years ago. The Italian demographer Massimo Livi-Bacci estimates that thirty thousand years ago, there may have been five hundred thousand humans, and by the beginning of the Holocene, just ten thousand years ago, there may have been five or six million.[29]

If we take just these last two figures, they suggest that human populations increased by about twelve times (or by an average of a quarter of a million every thousand years) in the last twenty thousand years of the Paleolithic period. On the reasonable assumption that each individual was using no less energy than before, that suggests that total human energy consumption also increased by about twelve times. Collective learning, over more than one hundred thousand years, had significantly increased human control over energy and resource flows in many different parts of the world.

Most of these increasing flows of energy supported population growth. Not much energy was spent on increasing complexity at the local level; as we have seen, human communities remained small and intimate. At the species level, though, there is no doubt that the spread of humans around the world represented an increase in complexity, because by ten thousand years ago, humans employed a much greater diversity of technologies and information than any other species on Earth, and they deployed it across much of the planet.

We have no evidence that more energy increased affluence. Some foragers may have lived pretty well. Indeed, the anthropologist Marshall Sahlins argued that in some environments, Paleolithic communities enjoyed varied diets, high levels of health, and large amounts of leisure time, which they could use for storytelling, for sleeping or relaxing, and for the marathon dances that seem to have bound most small communities together.[30] But there cannot have been significant differences in wealth, because foragers had no reason to accumulate goods when they could get most of what they needed from their surroundings. Besides, when you're regularly on the road, you want only the most valuable and portable of goods.

The coldest period of the most recent ice age, just over twenty thousand years ago, was followed by several thousand years of erratic warming until, starting about twelve thousand years ago, global temperatures settled into the warmer and more stable regime that dominated human history during the Holocene epoch. By the end of the last ice age, our alien scientists would already have been very interested in the strange events afoot on planet Earth. As climates got warmer, the behavior of humans would become even more striking. Quite suddenly (on paleontological scales), humans gained access to much larger flows of energy through farming, and these new flows of energy would allow a quantum leap in the complexity, diversity, size, and intricacy of human societies.

CHAPTER 8

Farming: Threshold 7

When Adam delved and Eve span, Who was then the gentleman? From the beginning all men by nature were created alike, and our bondage or servitude came in by the unjust oppression of naughty men. For if God would have had any bondmen from the beginning, he would have appointed who should be bond, and who free.

— JOHN BALL, SERMON PREACHED
DURING THE ENGLISH PEASANTS' REVOLT

Our ancestors lived as foragers for the first two hundred thousand years or more of our history. A constant trickle of innovations ensured that they would forage with increasing efficiency and in an increasing diversity of environments, until, by ten thousand years ago, at the end of the last ice age, humans were living in most parts of the world. In the past ten thousand years, human lifeways were transformed by a cascade of innovations that we describe as *farming* or *agriculture*.

Farming was a mega-innovation, a bit like photosynthesis or multicellularity. It set human history off on new and more dynamic pathways by helping our ancestors tap into larger flows of resources and energy that allowed them to do more things and create new forms of wealth. Like a gold rush, the bonanza of energy would generate a frenzy of change. Eventually it would transform the human relationship to the biosphere because, as farming societies grew, they supported much larger populations and evolved many more moving parts than foraging societies. More energy, resources, and people and more links between communities generated positive feedback cycles that accelerated change. For all these reasons, farming counts as our seventh threshold of increasing complexity.

The potential for transformative innovations had existed since collective learning first took off, but now that potential began to be realized as a result of three main Goldilocks conditions: new technologies (and increasing understanding of environments generated through collective learning), increasing population pressure, and the warmer climates of the Holocene epoch.

What Is Agriculture?

As human communities got better at collecting and managing information about their environments, they gathered and hunted with increasing understanding and skill, and their impacts on surrounding plants, animals, and landscapes grew. Fire-stick farming, for example, transformed vast areas, as it increased the production of plants and animals that were useful for humans. When Captain Cook and his crew sailed north along the east coast of Australia in 1770, they did not see wilderness. They saw distant spirals of smoke as Australians fired the land, and they saw landscapes as altered by human activity as the country gardens of their English homeland. Australia's megafauna were long gone. The fire-loving eucalyptus that now dominated so many Australian landscapes were there because of thousands of years of fire-stick farming.

Farmers, like foragers, used information accumulated over thousands of years. But they used it in new ways that would take human manipulation of the environment to an entirely new level.

The basic principle of farming is simplicity itself. Farmers use their environmental knowledge to increase the production of those plants and animals they find most useful and to reduce the production of those they can't use. Farmers weeded and watered the land to help grow the plants they wanted, such as wheat and rice, and fenced in animals they valued, such as sheep and goats, but they removed weeds and shooed away or killed animals they didn't like, such as snakes and rats. These activities changed entire landscapes, and plants and animals responded to these new environments, as they respond to all environmental changes, by adapting genetically, by evolving. That is why new breeds of plants and animals began to appear as farmers altered their surroundings. The species that flourished best were those that pleased humans, because those were the species humans looked after most carefully. More nutritious plants, such as domesticated wheat and rice, evolved, as did more helpful animals, such as domesticated dogs, horses, cattle, and sheep. Domesticated animals helped hunters, carried and hauled people and goods, or provided wool or milk. When slaughtered, they provided meat, skins, bones, and sinews.

Farmers found that transforming their environments was hard work. But in return for their chopping, plowing, weeding, draining, and fencing, they got a lot more energy and resources from the land, rivers, and forests that surrounded them, because the species they valued flourished spectacularly. That allowed the first farmers to tap more of the photosynthetic energy flowing through the biosphere. The total flow of photosynthetic energy did not necessarily increase, of course. It may even have declined as farmers removed high-productivity plants such as trees. But for farmers, the important thing was that they could now tap a larger share of the existing flows.

Farming gave farmers more than just food, wood, and fibers. It also gave them indirect access to new flows of energy. For example, humans cannot eat grass, but horses and oxen can, so farmers who let horses and oxen graze and then used them for riding or haulage or killed and ate them were tapping into the large flows of photosynthetic energy through grasslands. That makes quite a difference. A human can deliver at most about 75 watts of energy, while a horse or ox can deliver up to ten times as much. All that extra energy could be used to plow the land more deeply than handheld hoes could, or to cart goods or carry people. Farmers could also increase the production of plants and animals that had other uses besides food, such as flax and cotton, which could be used to make textiles. Or they could plant trees and use the wood to build homes, farms, barns, and fences, or burn it to cook their food and warm their houses.

Put simply, farming was an energy and resource grab by a single, very resourceful species with access to increasing amounts of information about how to exploit its environment. Through the magic of collective learning, humans had discovered how to increase their share of the energy and resources flowing through the biosphere by diverting more and more of those flows to human uses, just as humans would eventually channel major rivers onto their own fields and into their own cities.

To a biologist, farming looks like a form of symbiosis: an intimate and mutually beneficial relationship between distinct species. Foragers used and knew about hundreds of different species of plants, animals, and insects, but farmers focused on a small number of favored species, so they developed exceptionally intimate relationships with them. Intense symbiotic relationships often lead to changes in the behavior and the genetic makeup of both species. Modern honey ants "domesticate" aphids. They protect the aphids, provide them with food, and help them reproduce. Now the aphids have changed so much that they can no longer survive on their own. They pay for food and protection by supplying the ants with honey when the ants stroke them gently. More familiar and more important to us is the relationship between plants and bees. Bees get nectar, and the flowers reproduce more reliably because the bees carry their pollen from flower to flower. Kill off too many bees, and the grain harvest that feeds billions of humans today would be in serious trouble.

The favored species on which farmers lavished so much care and work (the domesticates) gained little in quality of life. But they did well demographically. Their numbers soared, while the number of wild animals (the animals farmers were *not* interested in) plummeted. In the year 2000, the total biomass of all wild land mammals was about one-twenty-fourth that of domesticated land mammals.[1]

Symbiosis changes all species involved, as they coevolve. Compare a modern ear of corn with teosinte, its bedraggled wild ancestor. Or compare a wild mouflon sheep with its modern, domesticated relative. The domesticated animal looks almost as if it had evolved to please humans. It is docile (some might say, unkindly, that it is stupider than its country cousins), it produces more wool than it needs, its meat is tasty for humans, and it cannot survive without human protection. Demographically, this is a surprisingly successful evolutionary strategy. Today there are more than one billion domestic sheep but just a few remnant populations of mouflon.

Humans changed, too, but in different ways. Most of their adjustments were cultural rather than genetic. Humans *have* changed genetically as a result of farming. For example, if you're descended from people who once herded cattle and consumed cow's or mare's milk, you will probably be able to digest their milk even as an adult because you can keep producing lactase, the enzyme that digests lactose (milk sugar). Hunter-gatherers consumed only breast milk till about four years of age, and after childhood, they no longer needed to produce lactase. But where cow's or mare's milk became a major food source, humans began to produce lactase into adulthood—a genetic mutation had occurred.

For the most part, though, humans adapted to the symbiotic

relationships of farming not with genetic changes but with new behaviors: technological, social, and cultural innovations accumulated through collective learning. They developed new ways of working the land, woodlands, and rivers. And as they did that, they had to learn new ways of collaborating and living together. Cultural change happens much faster than genetic change, and this explains why farming transformed human lifeways within just a few generations.

The History and Geography of Early Farming

It took humans one or two hundred thousand years to adapt their foraging technologies to the many different environments of planet Earth. Farming spread around the world in less than ten thousand years, as farmers adjusted their farming methods to different species, soils, and climates. Today we can trace the spread of farming the same way we can trace the spread of a disease from several different infection centers.

Farming did not expand evenly or smoothly. It spread fast in some regions, slowly in others, and hardly at all in still others, and these differences would have a huge impact on the geography of human history. By the time farming got going, humans were so widely dispersed that events in one part of the world had little impact elsewhere. Major changes took place community by community and spread first through local networks. Over time, ideas moved over larger distances, but, until five hundred years ago, there were some fundamental barriers to the movement of people, ideas, and technologies, including farming. Rising sea levels after the end of the last ice age severed links between Eurasia and the Americas, and there was hardly any communication between Eurasia and Australasia or with the islands of the western Pacific, some of which were settled as early as thirty thousand years ago. In effect, humans now lived in a number of separate world islands or zones. Within these zones, human history played out almost as if the inhabitants were living on different planets.

The largest and oldest world zone was Afro-Eurasia. This is where humans had evolved, and because there was a land bridge between Africa and Eurasia, ideas, people, and goods could move in relays over vast distances. The next-oldest world zone was Australia, first settled about sixty thousand years ago. The Australasian world zone was connected to Papua New Guinea and Tasmania during the last ice age but had the most tenuous of connections to Eurasia. The third-largest world zone, in the Americas, was settled at least by fifteen thousand years ago but was largely cut off from Eurasia when the Bering Strait was flooded at the end of the last ice age. In recent millennia, a fourth zone would emerge in the Pacific. Western islands such as the Solomons may have been settled as early as forty thousand years ago, but islands farther to the east and south (including New Zealand, Hawaii, and Easter Island) were settled during a remarkable series of seaborne migrations that began just thirty-five hundred years ago.

The existence of different world zones set up a fascinating natural experiment because, as we look back, we can watch how human history played out in different arenas.[2] There were important similarities in the histories of the world zones. Everywhere, collective learning generated new technologies, new social relationships, and new cultural traditions. But it did so at various speeds, and that meant that farming evolved in distinct ways, creating very different regional histories. These differences would prove immensely important when the world zones were reconnected after 1500.

Farming appeared first in the Afro-Eurasian world zone, and that is where it would spread farthest and have the largest impact. It also emerged quite early in Papua New Guinea. Eventually it would flourish in the Americas. Elsewhere, though many communities explored some forms of farming, its impact was less transformative.

By fourteen thousand years ago, foragers had spread to all the different world zones, and some, particularly in the Afro-Eurasian world zone, began to settle down and rearrange their surroundings. Five thousand years later, farming villages could be found at the hinge point of the African and Eurasian continents, along the river Nile and in the arc of highlands on the eastern shores of the Mediterranean known as the Fertile Crescent. Two thousand years later, in a quite different region, farming villages appeared in the highlands of Papua New Guinea. By four thousand years ago, you could have found farming communities in many parts of Africa and Europe, in much of southern, southeastern, and eastern Asia, and in the American world zone. By then it is probable that most humans depended on farming, because farming supported bigger populations than foraging did. But large regions of the world, including Australia, the Pacific, and many areas of the Americas and Afro-Eurasia, were still inhabited by thinly scattered communities of nomadic foragers, though even here, we sometimes see small steps toward farming.

Farming, or near-farming, evolved quite independently in different parts of the world. It was not a one-off invention. That suggests something very important: as independent human communities accumulated more technological and ecological knowledge, there was a high probability, wherever they were, that they would eventually use the knowledge they had accumulated as foragers to develop farming techniques. But they were likely to do so only if they needed the extra resources that farming could provide because, after all, farming was hard work and it meant changing a community's entire way of life.

Why Did Humans Take Up Farming? Crossing Threshold 7

At the end of the last ice age, two worldwide changes coincided to create a small number of regions in which farming began to look tempting. First, climates began to get warmer and wetter around the globe; second, foragers now occupied so much of the Earth that some regions were beginning to feel overpopulated. Both changes nudged humans toward farming. Because these changes were felt to some degree in different regions in all of the world zones, they help explain the strange fact that farming appeared within just a few thousand years in parts of the world that had no contact with one another.

Climates began to warm, erratically, about twenty thousand years ago, and by thirteen thousand years ago, average global temperatures were similar to today. Then, during the cold snap known as the Younger Dryas period, they fell sharply for at least a thousand years, after which they rose again. For about ten thousand years, climates have been unusually stable. Warmer, wetter climates and exceptional climatic stability made farming more viable than it had been for at least one hundred thousand years, providing the Goldilocks conditions for the entire agrarian era. Graphs of average world temperatures over the past sixty thousand years show clearly the remarkable climatic stability of the past ten thousand years, even though variations were greater away from the tropics.

The warmer, wetter climates of the early Holocene created a few regions of abundant and diverse plant life that formed rich "Gardens of Eden" for local foragers. In some of these regions, resources were so abundant that foragers could settle down in permanent communities or villages. Recently, nine-thousand-year-old circular stone houses have been found on the Dampier Archipelago, off the coast of Western Australia.[3] Similar changes have been studied most closely in the Fertile Crescent, on the eastern shores of the Mediterranean. Here, from fourteen thousand years ago, communities known to archaeologists as Natufians began to live in permanent villages with hundreds of people. They harvested wild grains using sickles made of sharpened flints embedded in the jawbones of asses. They kept gazelles in pens. And they built houses and buried their dead in cemeteries. They were not yet farming—pollen found at these sites belongs to wild grains. But they were sedentary and they lived in villages. Archaeologists describe such communities as "affluent foragers."

Population pressure may also have encouraged the Natufians to become more sedentary. There are a lot of Natufian settlements, which suggests that populations were growing fast in the Fertile Crescent. That is not surprising, because the Fertile Crescent lay across major migration routes between Africa and Eurasia, which may have funneled in new arrivals.

Settling down encouraged further population growth in several different ways. Foragers, who were well aware of how few people the land could support, often tried to limit population growth. However, in villages, infants no longer had to be carried and they could eventually be put to work. That changed attitudes to families, to children, and to gender roles. In villages, having lots of children provided plenty of labor for the household as well as protection and care for the old. That's why, in most sedentary communities, women were expected to bear as many children as they could, partly because they knew that perhaps half would die before they reached adulthood. Such attitudes sharpened differences in gender roles and ensured that most women's lives would be dominated by the bearing and rearing of children throughout the agrarian era of human history. The same rules explain why, within a few generations, many villages of affluent foragers faced the challenge of overpopulation.[4]

As populations grew, the Natufians had to extract more resources from the land. That meant grooming the land more carefully, and eventually it meant taking up some form of farming. The Natufians were falling into a honey trap. They had built their first villages in what seemed like an ecological paradise, but within just a few generations, they faced a new population crisis, and because neighboring communities were also growing fast, they could not just use more land. Instead, they had to use whatever tricks they knew of to increase the productivity of the land they already had. These pressures pushed them, probably reluctantly, into the tough life of farmers, and as they learned what it meant to be farmers, they forgot what it meant to be foragers. As always with collective learning, the accumulation of new knowledge eclipsed ancient knowledge and insights. Similar pressures would transform foraging communities in many different parts of the world as populations grew.[5]

Some of the best evidence for the transition from affluent foraging to farming comes from Abu Hureyra in the north of modern Syria, near the Euphrates Valley. The site was discovered in the early 1970s and excavated for just two seasons before it was flooded by the building of a dam. The earliest levels consisted of a cluster of round houses typical of Natufian foragers and dating to about thirteen thousand years ago. Their inhabitants hunted gazelles and wild asses and gathered a wide variety of foodstuffs, including nuts and fruit and wild grains. As climates deteriorated during the thousand-year-long cold snap known as the Younger Dryas, warm-weather fruits vanished, and villagers began to rely on more hardy grains, even though these were more difficult to gather and process. Eventually, they relied on domesticated varieties of the cold-adapted grain rye, so, in Abu Hureyra, at least, it seems to have been climate change that turned foragers into farmers. Toward the end of the cold spell, the site was abandoned for many centuries; it was reoccupied almost eleven thousand years ago. Now a substantial village appeared, with hundreds of rectangular mud-brick houses and several thousand inhabitants who cultivated domesticated grains and hunted wild gazelles and sheep. Then, quite rapidly, the number of sheep bones increased, a sure sign that sheep were now fully domesticated. The human remains reveal how tough life could be for the first farmers. All have heavily worn teeth from a diet dominated by grains, though tooth wear diminishes with the appearance of pottery, which made it possible to process grains into gruels. Women's bones show clear evidence of wear from long hours of rocking back and forth on their knees as they ground grain.[6]

We can be pretty sure that the first farmers took up farming reluctantly, because living standards seem to have declined in early agrarian villages. The skeletons found in early farming villages in the Fertile Crescent are usually shorter than those of neighboring foragers, which suggests that their diets were less varied. Though farmers could produce more food, they were also more likely to starve, because, unlike foragers, they relied on a small number of staple crops, and if those crops failed, they were in serious trouble. The bones of early farmers show evidence of vitamin deficiencies, probably caused by regular periods of starvation between harvests. They also show signs of stress, associated, perhaps, with the intensive labor required for plowing, harvesting crops, felling trees, maintaining buildings and fences, and grinding grains. Villages also produced refuse, which attracted vermin, and their populations were large enough to spread diseases that could not have survived in smaller, more nomadic foraging communities. All this evidence of declining health suggests that the first farmers were pushed into the complex and increasingly interconnected farming lifeway rather than pulled by its advantages.

How did they know how to get more crops from the same amount of land? How, in fact, did they know how to farm? This is where the power of collective learning is most apparent. Most other species faced with similar ecological crises would have hit a demographic brick wall. That wall explains the familiar S-shaped curve of population growth in most types of organisms: a new species multiplies until it is extracting all the food energy in its niche, after which individuals starve, fertility falls, and population growth plateaus. Humans had more options because they had more information. Much of that information had not been needed before. It was potential knowledge, like potential energy—knowledge held in reserve that could be activated if and when it was needed. Modern foragers have a lot of potential knowledge that can be activated in a crisis, and Natufians surely had similar forms of knowledge. They knew that the plants they liked grew better if you irrigated them and if you removed competitors by weeding. In Australia in recent centuries, foraging communities introduced more intensive technologies, such as harvesting grains (using sickles made from stone blades with handles covered in fur, in northern Australia), grinding seeds, or breeding eels in specially built systems of small canals.[7] But most of the time, foragers don't bother with these technologies, because they aren't needed, and they require a lot of extra work. In regions such as the Fertile Crescent, the climatic and demographic changes of the early Holocene provided both the opportunity and the motivation to use these reserve technologies and to use them more or less continuously. That is what turned foragers into farmers.

In summary, warmer climates made village life and farming possible in a few favored regions, population pressure sometimes made it necessary, and the reserve knowledge accumulated by foragers over many millennia provided the start-up technologies for the first farmers.

The geography of early farming was shaped by the happenstance of plate tectonics and the types of plants and animals that had evolved in particular regions. Some plants and animals could be domesticated quite easily. Others could not. Foragers were attracted to regions like the Fertile Crescent, areas that had plants and animals that were ripe for domestication.[8] Foragers surely auditioned many different species as potential domesticates. Among the most attractive plants were those that built up rich stores of nutrition for their seeds, such as fruit trees. Even better were seasonal plants with tubers or fat seeds that stored up nutritional goodies to help humans survive dry periods. Wheat and rice, if harvested at their peak, provided such concentrated sources of nutrition that they were worth the huge effort required to plant, protect, water, harvest, and store them.[9]

Animals, too, varied in their usefulness. Zebras were too ornery to be tamed. Lions and tigers were too dangerous and not particularly tasty. But herd animals such as goats, cattle, and horses were easier to control, particularly if humans could stand in for the leader of the herd. If the animals were grass-eaters, they could turn grass into meat, milk, fibers, and power, enabling humans to exploit the world's vast grasslands. And their meat was usually tasty and nutritious. But by the time agri-culture began to spread, large domesticable herbivores could be found only in Afro-Eurasia. As we have seen, most megafauna (with the partial exception of South American camelids such as llamas) had been driven to extinction in Australasia and the Americas, probably soon after the arrival of humans. This may help explain why agriculture flourished earlier and spread more widely in the Afro-Eurasian world zone than in the other world zones.

The Early Agrarian Era: Farming Spreads Around the World

After appearing in several core zones, farming villages multi-plied and spread, as farmers honed their skills, learned new ways of increasing production, and took farming into new regions.

Major rivers that had laid down fertile alluvial soils over thou-sands of years, such as the Tigris and Euphrates, the Yellow (Huang He) and Yangtze Rivers in China, and the Indus and Ganges Rivers in the Indian subcontinent, lured increasing num-bers of farmers. Farming villages appeared in the Fertile Cres-cent and the Nile basin perhaps eleven thousand years ago, then along the Yangtze and Yellow Rivers within a millennium or two.

By six or seven thousand years ago, food crops such as taro were being cultivated in the highlands of Papua New Guinea. Between five thousand and four thousand years ago, there were farming villages in the Indus Valley and in West Africa. Farmers also appeared now in the American world zone: along the Mississippi River, in parts of modern Mexico and Central America, and in the Andes, whose mountains provided diverse environments and a wide range of potential domesticates.

There was nothing automatic about the spread of farming from the core regions where it first appeared. For example, it did not spread from the Papua New Guinean highlands into the coastal lowlands, where highland crops such as taro and yams didn't flourish as well.

As population pressure drove migrants into new environments, they had to adapt their farming techniques, and sometimes they had to wait until their domesticates had evolved new varieties. From the Fertile Crescent, farming extended into Central Asia, Turkey, and then into the Balkans, Eastern Europe, and Western Europe between eight thousand and four thousand years ago. As farming spread into the cooler, more forested regions of Europe, with their different soils, growing seasons, and pests, both farmers and their crops had to adapt. In central and northern Europe, farmers developed new varieties of grains. In forested regions, they took up shifting, or swidden, agriculture, a sort of nomadic farming. Swidden farmers burned and cut down trees, then farmed the ashy soil between tree stumps. After a few years, when the soil lost fertility, they moved on. In the Indus Valley, farming flourished four thousand years ago, receded, then advanced again, starting about three thousand years ago, along the Indus and the Ganges Rivers and into other parts of the Indian subcontinent. In Africa, cattle herders flourished in the Sahara (which was wetter and more productive than today) by five thousand years ago and maybe much earlier. By three thousand years ago, farming was well established in West

Africa. From there, it spread into central and southern Africa. In the Americas, too, farmers had to adapt to new conditions; for example, distinct varieties of maize evolved in Mesoamerica and along the Mississippi River.

As farming communities multiplied, the pace of change accelerated, because farming, and the many changes it brought with it, spread faster than foraging. Why farming increased so rapidly is not immediately obvious, because the farming life could be tough, and that's why foragers survived, often alongside farmers, for many millennia. In some regions, such as Siberia and Australia, the disadvantages of farming outweighed the advantages, and foragers flourished until modern times. Nevertheless, in regions suitable for farming, regions that could be *made* suitable for farming, or regions where rapid population growth pressed hard on available resources, farming communities had many advantages over their foraging neighbors. Even swidden agriculture could support around twenty to thirty people per square kilometer. That was about one hundred times the population densities typical of foragers in similar environments.[10] When push came to shove, this meant that farming communities could usually mobilize more people and resources than foragers could. They could swamp them demographically and, if necessary, defeat them militarily. That is why, by perhaps as early as five thousand years ago, most humans depended on farming, and farming communities and those they supported were beginning to dominate human history.

As farmers spread, they transformed their surroundings. Everywhere, farmers cut back forests, built villages, plowed up the land, drove off pests, and dug up weeds. By its very nature, farming required a manipulative attitude to the environment. While foragers normally thought of themselves as embedded within the biosphere, farmers saw the environment as something to be managed, cultivated, exploited, improved, and even conquered. And while collective learning gave farmers the knowledge they needed to manipulate their environments, farming gave them the food and energy flows they needed to multiply and to transform their surroundings over larger and larger areas and with increasing power and virtuosity.

Collective learning and new energy flows — these drove the turbulent historical dynamism of the agrarian era, accounting for a disruptive changeability that had not been seen in the Paleolithic.

How Farming Transformed Human History

For perhaps five thousand years after the end of the last ice age, the agrarian era of human history was dominated by farming villages. These were the megalopolises of their era, the most complex, populous, and powerful communities on Earth. As farming spread and populations grew, villages multiplied until they became the communities in which most humans lived. If you were a human in the agrarian era, you were probably a farmer or lived in a community of farmers.

Such dense communities were a new phenomenon in human history. By modern standards, farming villages may look simple. But by Paleolithic standards, they were social, political, and cultural juggernauts. They required not just new technologies but also new social and ethical rules, new ideas about how to live together, how to avoid conflicts, and how to divvy up the community's wealth. If British anthropologist and evolutionary psychologist Robin Dunbar is right that evolution equipped human brains to cope with groups of no more than 150 individuals, it follows that communities much larger than this would need new social technologies to hold them together.

During the first half of the agrarian era of human history, most farming villages were independent communities with limited ties to neighboring villages and small enough to be held together through traditional kinship rules. Though exchanges of people, goods, and ideas between villages were increasingly important, there were not yet any states, empires, cities, or armies. The huge, complex societies that have dominated the last five thousand years of human history appeared only after farming had spread far and fast enough to create a critical mass of people, resources, and new technologies. But the roots of agrarian civilizations can be found in the village communities of the early agrarian era.

We have already seen that foraging societies contained reserves of potential knowledge of many different kinds, including information about how to manage large groups of people. The *potential* for increasing social complexity, for large-scale political, economic, and military networks, and for the huge buildings that we find in all agrarian civilizations was already present in foraging and early farming communities.

Göbekli Tepe, in southern Anatolia, offers a spectacular illustration of the intellectual and technological potential lurking within early foraging and farming communities. Göbekli Tepe was occupied first during the era of Natufian villages and then periodically between twelve thousand and nine thousand years ago.[11] It contains twenty stone circles with about two hundred beautifully carved stone pillars, some of them well over five meters tall and weighing up to twenty tons. Many have strange bas-relief images of clawed or beaked birds or animals. There are no domestic buildings, and, curiously, many of the pillars were ritually buried. Archaeologists have also found hints of beer brewing on the site, and that, too, may hint at ritual activities (as well as bacchanals). This suggests that Göbekli Tepe, like Stonehenge in England or Chaco Canyon in New Mexico, was a ritual center for surrounding communities, an early equivalent, perhaps, of the Olympic Games or the United Nations. It may also have functioned as an observatory. The huge effort that went into building Göbekli Tepe's stone circles suggests the importance of diplomatic and technological links between different communities in an era of rapidly increasing populations. The size of the pillars, the precision and beauty of the carving, and the fact that hundreds of people must have been employed to carve and move the large stone blocks point to a new scale and complexity of social organization. This is surprising because it is likely that those who built the earliest of these structures were not yet even true farmers but were, like the Natufians, sedentary or affluent foragers.

Traditional rules of kinship were challenged as villages and networks of villages grew larger.[12] As early farming villages expanded, built new links with neighbors, and sometimes turned into small towns, traditional rules of kinship and family had to be modified or supplemented with new rules about property, rights, ranking, and power. The traditional social modules of one or two hundred people had to be linked into larger networks that were, inevitably, hierarchical. Everywhere, as farming spread, we begin to see new and more hierarchical structures that overlay village communities organized by traditional kinship rules.

One way to track relationships and rankings in a village of a thousand people is to use traditional kinship rules but project them back in time. Here's how it might have worked: If your parents and grandparents and great-grandparents were all descended from the eldest children in each generation, then you could claim the seniority of an elder child for yourself and your whole family. Mechanisms like this made it possible to rank whole families and lineages by seniority. Here we see the beginnings of classes and castes. But talent mattered, too. As people lived closer together in large villages, disputes increased over land rights, inheritance, assaults, or damage to property, like the collisions between protons in the contracting clumps of matter from which the first stars formed. But settling disputes in a large village was very different from sorting out a family quarrel.

Mediators or judges needed delicacy, tact, intelligence, and experience. And sometimes they needed to be able to impose their will by force.

Modern studies of small-scale village societies show how such problems can generate simple forms of leadership, as individuals who are known as particularly generous or forceful, particularly knowledgeable about traditions and the law, particularly pious, or particularly skilled in battle are granted a modest degree of authority over other villagers. If they are socially and politically adept, they may become "big men," leaders known for their generosity and their leadership and organizational skills. Ranks based on lineage or ability laid the foundations for divisions by class and caste. The outlines of imperial power were already prefigured in the feasts and fights of ancient villages.

With more people and more exchanges, the machinery of collective learning operated with increasing synergy and power. Many innovations offered incremental improvements to farming in different areas, and some innovations were game changers. Two particularly important innovations were the domestication of large animals and the emergence of large-scale irrigation.

Animals were probably domesticated at the same time as the first plants were. Dogs may even have been domesticated in foraging societies and used to help with the hunt, as guards, or even to keep people warm during the winter. But at first, animal domestication was inefficient. Animals were kept penned and fed, at considerable cost, until they were slaughtered for their meat, hides, bones, and sinews. By six or seven thousand years ago, particularly in regions with wide areas of grasslands that could support large herds of livestock, farmers and herders developed ways of exploiting domesticated animals before they killed them. They began to milk cows, mares, goats, and sheep; they sheared sheep and goats; and they rode horses or hitched them to carts. The archaeologist Andrew Sherratt described these new techniques as the "secondary products revolution," because humans had learned to use both the primary products of domesticated animals (the resources they yielded when killed) and their secondary products (the energy and resources they could supply while alive). Until modern times, these powerful technologies were limited to the Afro-Eurasian world zone because in the Americas, the killing off of many species of megafauna left too few potential animal domesticates. In some regions of Afro-Eurasia, however, such as Central Asia, the Middle East, and North Africa, the gains in productivity from secondary products were so great that entire communities began to live off their livestock, following them from grassland to grassland, living in tents, and returning to a nomadic way of life. We call such people *pastoral nomads*. Their mobility made pastoral nomads perfect connectors between distant regions, and eventually, they would carry ideas, technologies, people, goods, and even diseases right across Afro-Eurasia through the so-called Silk Roads.

Large-scale irrigation was equally transformative. In Mesopotamia, population pressure drove more and more farmers from the well-watered highlands of the Fertile Crescent into the arid southern lands at the heart of modern Iraq, through which flowed the region's two great rivers, the Tigris and the Euphrates. Here, there was so little rainfall that if you wanted to farm, you had to divert water from the rivers. At first, farmers used simple ditches that they dug themselves. Eventually, though, whole communities collaborated to build and maintain elaborate systems of canals and dikes. The largest of these systems demanded thousands of workers and a lot of leadership and coordination. But the payoff was huge in a region whose soils had been enriched for millennia by flooding from the major rivers. Farming advanced by leaps and bounds in regions suitable for irrigation, including North India, China, Southeast Asia, and, eventually, some regions in the Americas. Irrigation farming supported larger populations, but it also required increasing social cooperation, so it tended to bind farming villages into larger social and political networks.

Populations rose fast as farming methods improved and farming spread. It had taken at least one hundred thousand years for human populations to reach five million, at the end of the last ice age. By five thousand years ago, human numbers had quadrupled, rising to about twenty million. By two thousand years ago, there were two hundred million humans, forty times the number at the end of the last ice age.

But population growth was never steady. Everywhere, it was interrupted by catastrophes. Disease, famine, war, and death—the Four Horsemen of the Apocalypse—flourished in the agrarian era. As mentioned earlier, unlike nomadic camps, villages accumulated waste and attracted vermin, so diseases spread fast. Where new diseases appeared—infections for which people had no immunity, such as smallpox—it was not uncommon for half the population to die. Farmers were also more vulnerable to famine than foragers were, because they relied on so few crops. When food began to run out, weeds, acorns, and tree bark could support people only for so long, and the very young and very old suffered most and died first. As populations grew, villages fought over land, water, and other resources. Their battles summoned the Third Horseman, war, who could be even more ruinous than disease and famine and often worked alongside them. Humans had always fought, but in farming societies more people were involved, and weapons became more lethal as fighters acquired metal spears, chariots, and siege engines. The Fourth Horseman, death, rode behind the other three.

For better or worse, human history had entered a more dynamic era in which change was the one constant. As human communities grew in numbers, size, and complexity, they laid the foundations for the agrarian civilizations that have dominated the past five thousand years of human history.

CHAPTER 9

Agrarian Civilizations

In those days the dwellings of Agade were filled
with gold,
its bright-shining houses were filled with silver,
into its granaries were brought copper, tin, slabs of
lapis lazuli, its silos bulged . . .
its quays where the boats docked were all bustle . . .
its walls reached skyward like a mountain . . .
the gates — like the Tigris emptying its water
into the sea,
holy Inanna opened its gates.

— SUMERIAN POEM, TRANSLATED BY S. N. KRAMER

Farming villages and their populations provided most of the human and material resources for the agrarian civilizations that have dominated the past five thousand years of human history. Look behind the imperial armies and cities, the temples and pyramids, the trade caravans and shipping fleets, the literature and art, the philosophies and religions of agrarian civilizations, and you will find, in the background, often far from the heartlands, thousands of farming communities, as well as a large and even poorer population of vagrants and the dispossessed, many of whom were slaves. People from these underclasses produced most of the grains and meat, many of the linens and silks, and much of the labor (both free and unfree) needed by the great cities. Their produce and labor paid for the causeways and palaces and temples and the silks, wines, and jewelry of the rich, while their men and horses served in the armies. Agrarian civilizations mobilized the human and material wealth and the energy produced by farming villages to build social structures much more awesome and complex than any earlier human communities. Like all living organisms, they mobilized information, too, because more information gave them access to more energy and more resources.

The appearance of agrarian civilizations represents another threshold of increasing complexity. However, agrarian civilizations were built on foundations created by the evolution of farming communities over several millennia, so we will treat their appearance not as an entirely new threshold but as a second phase of the threshold that gave us agriculture.

To understand the emergence of agrarian civilizations, we will focus not on the histories of particular civilizations but rather on the questions we have posed throughout our modern origin story: What were the Goldilocks conditions for this new form of complexity? What were the new emergent properties of agrarian civilizations? And what were the flows of energy that sustained those new properties?

Surpluses, Hierarchies, and a Division of Labor

Despite famines, diseases, and wars, farming villages multiplied and spread throughout the Holocene because most years they produced more than they needed. They turned energy from sunlight into surplus wealth. This is very different from foraging societies, which stored knowledge but rarely felt any need to store surplus goods because the food and raw materials they needed were all around them. Why work as a farmer, asked modern foragers in the Kalahari Desert, when there are so many mongongo nuts to eat?[1] In foraging societies, the slow accumulation of knowledge encouraged migration into new environments rather than the accumulation of material goods. By contrast, farming societies *had* to store goods, and in large quantities, because many plants and animals were harvested over just a few weeks but eaten or processed over a year or more. So all farming communities had households, barns, sheds, and fields full of produce waiting to be consumed.

As productivity increased, surpluses began to exceed the annual needs of those who produced them. Surplus people, surplus food, surplus goods, and surplus energy represented new forms of wealth, which raised the question: Who was going to control (and enjoy) this wealth? Over time, surplus wealth would be mobilized by small but powerful minorities, and the structures they built to mobilize surplus wealth, often using crude forms of coercion, would form the muscles and sinews of agrarian civilizations.

Surplus wealth meant surplus people. As productivity rose, not everyone needed to farm, so new social roles appeared. Many people became vagrants or slaves, but other nonfarmers ended up controlling much of society's surplus wealth because they could specialize in useful social roles. They could become full-time priests or potters or soldiers or philosophers or rulers. Specialists became expert at their limited roles. But the division of labor also created new forms of dependence. As social roles multiplied, human societies, like the first metazoans, became more networked, more differentiated, more interdependent, and more complex. And new linking structures arose, the social equivalents of skeletons, muscles, and nervous systems.

Specialists were generally more dependent on the linking structures than farmers, who could usually feed themselves. Archaeologists can track the evolution of a division of labor. In Mesopotamia, pottery provides the classic case study. The earliest Mesopotamian pots are simple and idiosyncratic, and most were probably made in ordinary farming households. But from about six thousand years ago, we find special workshops with potter's wheels. Potters produced large quantities of standardized bowls, plates, and jugs and sold them over wide areas. These wares look like the work of full-time professionals who had invested in specialist equipment and long apprenticeships. Specialization encouraged new skills and techniques, so it was both a measure and a driver of technological change. For example, potters needed furnaces to fire their pots, and over time they built more efficient furnaces that operated at higher temperatures and yielded better finishes. But better furnaces were just what was needed to separate copper, tin, or iron from the ores in which they were embedded so the metals could be molded, bent, or hammered into household goods, ornaments, and weapons. Coppersmiths, goldsmiths, silversmiths, and blacksmiths all used technologies pioneered by professional potters.

As surpluses grew, specializations multiplied. Five thousand years ago, in the southern Mesopotamian city of Uruk, someone compiled a list of a hundred different special roles, the Standard Professions List. It was obviously important and widely known, because similar lists were copied by trainee scribes for many centuries. Organized hierarchically, the list includes kings and courtiers, priests, tax collectors and scribes, silver workers and potters, and even entertainers such as snake charmers. Potters and snake charmers, unlike farmers, did not produce food or leather or fibers, so they fed and clothed themselves and their families by exchanging their products and services for food and other necessities. This is why trade and markets and accounting devices such as coins and writing were as vital to complex societies as arteries and veins are to human bodies. They made it possible to transfer objects and the energy flows they represented, from person to person and from group to group. Even the religious specialists we describe as priests had to trade their spiritual services for food and other necessities. Where we find temples, we also find donations and gifts.

The degree of specialization was limited by the productivity of agriculture and by the number of extra people that each farmer could feed. In most agrarian civilizations it took about ten farmers to support one nonfarmer. That is why most people had to farm. Even in the first cities, most people grew crops in their backyards or outside the city walls. But while farmers made up most of the population and provided most of society's resources, specialists became increasingly important as societies became more interdependent. Farmers began to buy trinkets or farm tools and found they had to deal with peddlers, tax collectors, landlords, and overseers. Various different specialists moved goods and resources between towns and cities, produced the coins used in markets and the metal plows and swords used by farmers and soldiers, kept the accounts, policed the laws, prayed to the gods on everyone's behalf, or organized and ruled others. Specialists provided the struts and bracing for agrarian civilizations. That's why, eventually, they ended up organizing and dominating the rest of society.

As specialization increased, so did inequality. The earliest farming communities were reasonably egalitarian, even when they exceeded the ancient community maximum of 150 to 200 people. The Neolithic town of Çatalhüyük (in modern-day Turkey) flourished eight to nine thousand years ago, and it shows little variation in the size of domestic dwellings, even though its population may have reached several thousand. Eventually, however, we start finding wealthy minorities, and more and more of them. To take one random example: There is a six-thousand-year-old burial site near Varna on the Black Sea that contains more than two hundred tombs. Many of the dead were buried with nothing or with just a few simple objects, but about 10 percent of the graves held much more; one contained over a

thousand objects, most of them made of gold, including brace-
lets, copper axes, and even a penis sheath.[2] This is a very familiar triangle of wealth, with an elite population of about 10
percent and one person at the very top, while most people lived
close to subsistence. When archaeologists find small children
buried with great wealth, they can be sure that there are not just
hierarchies but hierarchies that cross generations, because children could not have attained high status on their own. These
are signs of aristocracies and castes. Large building projects,
such as palaces and pyramids and ziggurats and temples, also
tell us that someone had the power to organize the labor of
many others.

As gradients of power and privilege steepened, new social
struts were needed to sustain them. Someone had to police markets, punish pickpockets and thieves, count tax payments, and
organize peasants, vagrants, and slaves into the work gangs that
built palaces and maintained canals. Complex societies also
needed religious specialists to ensure that their gods protected
them from disease and provided plenty of rainfall. When these
structures failed, everyone was affected, and that is why most of
the time, even those at the bottom of the heap usually obeyed
their overlords.

Anthropologists have studied the emergence of hierarchies
in modern small-scale societies, such as those of Melanesia in
the western Pacific. Here, powerful figures, known to anthropologists as big men or chiefs, built their power on respect and
the loyal support of family, allies, and followers. But their power
was always precarious. If they failed to distribute enough wealth
and privilege to maintain the loyalty of their followers, they
could swiftly lose their power, their wealth, and sometimes their
lives. Why follow someone who cannot coerce you and from
whom you receive no benefits?

Eventually, in larger societies, there appeared much more
powerful leaders; they ruled over hundreds of thousands of

people and controlled such huge flows of wealth that they and their allies could buy the muscle needed to impose their will by force when necessary. Indeed, the use of force to extract labor or produce or wealth became ubiquitous in agrarian civilizations. That is why slavery and forced labor were common in agrarian civilizations. And the methods used to extract wealth and labor from peasants show that their condition was often little better than that of slaves. A wonderful document from Egypt, written late in the second millennium BCE, gives some insight into the methods routinely used to make peasants surrender surplus resources.[3] The author, a scribe, explains why it is good to be a scribe. Think of the hard labor of a peasant, the long hours spent working the fields in the heat and cold or looking after livestock or mending farming equipment and buildings. And then imagine what can happen when the tax collectors turn up with armed bodyguards.

> One says [to the peasant]: "Give grain." "There is none"
> [he says]. He is beaten savagely. He is bound, thrown in
> the well, submerged head down. His wife is bound in his
> presence. His children are in fetters. His neighbors
> abandon them and flee.

There is surely some caricature here, but we have plenty of evidence that extortionate methods were used in all agrarian civilizations to maintain order and to extract labor and resources from the majority of the population.

We generally refer to power structures capable of exerting this sort of control over extensive areas as *states*. States emerged in societies that were populous and wealthy enough to have towns and cities as well as large numbers of farming villages and plenty of surplus labor that could staff and pay for armies and bureaucracies.

From Towns to Cities and Rulers: Mobilization and a New Trophic Level

As populations and surpluses grew, so did the size of the largest human communities. And communities as well as people began to specialize. Some villages grew and acquired new roles because they were near trade routes, controlled strategic river crossings, held markets that attracted buyers and sellers from other villages, or were near important religious sites. Çatalhüyük in southern Anatolia was surrounded by good farming land, but it also had obsidian, the hard volcanic glass used to make the finest and sharpest Neolithic blades. Its inhabitants may have traded obsidian as far as Mesopotamia. Jericho, one of the oldest sites of continuous settlement anywhere in the world, was first settled in Natufian times because it had a well that never ran dry. By nine thousand years ago, Jericho had evolved into a town of perhaps three thousand people.

As towns grew, some offered new services, jobs, and goods. More people were lured to them, and over time they acquired power over the villages and towns in their hinterlands. By five thousand years ago, some large towns had turned into cities, huge, diverse communities supported by surrounding towns and villages and with large concentrations of specialists. The diversity of skills, jobs, goods, and people found in cities explains why they became technological, commercial, and political dynamos in all agrarian civilizations and why they sucked people in from the surrounding countryside.

The appearance of cities and states marks a fundamental transformation in human societies.

Traditional states were very different from modern states. Above all, they lacked the communication technologies and bureaucracies that allow modern states to reach into the lives of all their citizens. Traditional rulers could exert immense force locally, but it could take weeks or months to send an order to outlying provinces and as much time again to learn the outcome. So, away from the major population centers, the power of rulers depended on loose, hierarchical networks of local lords, who often governed their own territories as more or less independent fiefdoms. Nevertheless, the first states were a new phenomenon in human history. They all assumed the right to mobilize wealth from farming communities, towns, and cities in return for some degree of protection. As the English political theorist Thomas Hobbes wrote in *Leviathan* (1651), the right to distribute resources "belongeth in all kinds of Common-wealth, to the Soveraign power. For where there is no Common-wealth, there is...a perpetual warre of every man against his neighbor." Traditional elites owed their power, in part, to the intrinsic weakness and isolation of traditional farming communities. As Karl Marx noted, peasants had no more unity than potatoes in a sack.[4] That made them vulnerable to predation, because even weak rulers could use small numbers of enforcers to impose their will, village by village. This uneven balance of power explains why, for many thousands of years, small groups of rulers and officials successfully dominated much larger populations of farmers.

The history of the first cities, the first states, and the first agrarian civilizations is best known in Sumer, in southern Mesopotamia. Here, a large cluster of cities emerged quite rapidly about fifty-five hundred years ago. The southern Mesopotamian city of Uruk is often described as the first city in human history. It was a port on the Euphrates River. Like most Mesopotamian cities, it depended on complex, well-managed irrigation systems fed by the major rivers. But it also bordered the swamps of the southern river delta. Indeed, it may have grown in a period of drying climates, which forced people from outlying villages to migrate into the cities with their well-maintained irrigation systems. Fifty-five hundred years ago, Uruk had a population of ten thousand people living on opposite banks of the river Euphrates. Two hundred years after that, it probably had as many as fifty thousand inhabitants living in an area of about two and a half square kilometers.[5] At some point, the river Euphrates shifted its course and began to run around the edge of the city.

A city of fifty thousand people may not sound impressive today. But in its time, Uruk was a monster, perhaps the largest settled community that had ever existed in human history. It had two big temple complexes. That means there must have been powerful priests or kings who could mobilize the labor of thousands of people, many of them slaves. Uruk had workshops that made objects of great beauty, and it had storehouses for grain and precious goods. Accounts from a few hundred years later give us some idea of what you might have seen if you had visited Uruk when it was the capital city of King Gilgamesh, the hero of the first written epic. There would have been large temple complexes and royal palaces. You would have seen gardens and narrow streets and alleyways with workshops, inns, and shrines. The city was surrounded by a wall of burned brick, and canals led to the harbor and nearby farmlands. In the epic of Gilgamesh, the king says: "One third of the whole is city, one third is garden, and one third is field, with the precinct of the goddess Ishtar." Archaeologists have found Uruk-style goods as far afield as Anatolia and Egypt, which suggests that Uruk's merchants were trading over a large area.

Sometime around five thousand years ago, the first writing appeared in Uruk, on clay tablets in the temples of Eanna. More complexity meant more information, and writing was the new technology that allowed the wealthy and powerful to keep track of the increasing resources and energy flows at their disposal. Almost all the earliest writing in Mesopotamia consists of inventories—so many cows and bulls, so many sheep, so many bales of linen, so many slaves. They tell us that we are now in a world of rapidly increasing inequality in which networks of rulers, aristocrats, and officials control flows of information and power that enable them to mobilize the energy and produce of large numbers of slaves, farmers, and artisans.

A wonderful artifact known as the Standard of Ur, exhibited in reconstructed form in the British Museum, gives us a vivid glimpse into the cities of southern Mesopotamia almost five thousand years ago. The Standard of Ur is a boxlike object that may have been part of a musical instrument or may have been carried in parades; its actual function is uncertain. On its sides are mosaic images made of shells from the Persian Gulf, lapis lazuli from Afghanistan, and red stones from India. One side shows the city of Ur at peace. A kinglike figure and wealthy lords sit at a banquet being serenaded by a singer with a lyre. The king and nobles are larger than the servants, an artistic convention that highlights their rank and importance. Lower panels show goods and livestock being brought to the city, perhaps for the feast. The surpluses produced by farmers are being pumped upward to be consumed by elite groups. The other side of the standard shows Ur at war and illustrates some of the forces that maintained these steep gradients of wealth and power. At the top is a figure who is larger than all the others and is surely a king. Below, we see troops, apparently in official uniforms, and military leaders riding chariots pulled by donkeys. Some seem to trample enemy soldiers, while others drag naked captives with clearly visible wounds.

The cities of southern Mesopotamia five thousand years ago represent the sort of society that would dominate the history of the next few thousand years. Expensive, well-equipped armies allowed rulers and the elites that supported them to repel foreign enemies and maintain the gradients of power and wealth on which their own power and wealth depended. Like the proton pumps that maintain an energy gradient across cell membranes, soldiers and the armed retinues of nobles maintained gradients of persuasion and coercion that pumped wealth from villages to towns to cities and governments. Images of these power hierarchies, with magnificently dressed kings and overlords menacing their enemies and subjects, appear in all agrarian civilizations.

Viewed ecologically, states and their rulers represent a new step in the food chain, a new trophic level. We have seen how energy from sunlight enters the biosphere through photosynthesis and travels from plants to herbivores to carnivores. And we have seen how most of that energy gets wasted at each trophic level, in a sort of garbage tax. That leaves much less energy to support the higher levels, which is why there are fewer lions than antelopes. Agriculture increased the resources available to humans, so states could add one more trophic level at the top of the hierarchy. Rulers and nobles and officials began to squeeze wealth from the labor and produce of peasants, who in turn got their energy and food from farming. States used these new flows of labor, produce, and energy to pay for the armies, bureaucracies, palaces, and goods that made them powerful and wealthy.

Thinking about such processes in ecological terms reminds us that wealth never really consists of things; it consists of control over the energy flows that make, move, mine, and transform things. Wealth is a sort of compressed sunlight, just as matter is really congealed energy. Mobilizing this compressed energy from the rest of the population, along with the flows of resources that it made possible, became the fundamental task for rulers and governments, and that task would shape all aspects of the evolution and history of agrarian civilizations.

Indeed, mobilization was more central to the work of traditional states than it is to that of modern states. Traditional rulers did not need to concern themselves too much with the education, health, or day-to-day lives of most of their subjects because peasants could generally support themselves. In fact, many peasants continued to live in independent villages well beyond the reach of states and empires, so where states did rule over peasants, their main task was to extract resources from them. And over time, rulers, officials, and nobles became increasingly skillful at the task. If they needed more resources to build palaces or roads, recruit new legions of soldiers, or pay for their own luxury goods, few traditional rulers opted for the modern strategy of investing in productivity-raising innovations. They were conservative technologically because change was so slow that innovations rarely yielded significant returns within a human lifetime and often disrupted existing flows of wealth. Rulers might invest in new weaponry or build roads, but for the most part, the challenge was to increase available resources with existing technologies through traditional forms of mobilization.

To increase their wealth and power, traditional rulers had three main options. The most foresighted encouraged peasants to plow up unfarmed land and urged merchants to seek out new commodities. But many sought more rapid gains by using two other riskier and more coercive strategies. They could press harder on their own populations, at the risk of popular uprisings or economic breakdown. Or they could gamble on taking wealth from neighboring states by sending in their armies. This was dangerous, but it often worked, and that is why most traditional elites were warlike. That also explains why, when rulers had statues made in their honor, they usually posed wearing armor and carrying weapons. This was, after all, a world in which resources were mobilized primarily through the threat of coercion and in which the ability to mobilize and inflict violence was widely admired. If you were a king, taking resources from your neighbors was one of the most important ways of growing your economy. And if you succeeded (think Alexander the Great), you would probably be admired, no matter how much misery you caused.

The central role of mobilization is apparent from the manuals of statecraft that many traditional rulers produced. One of the richest examples is an Indian manual of statecraft, the *Arthashastra*. It was probably written slightly less than two thousand years ago, but it assembled the accumulated experience of many earlier manuals. Powerful states had emerged in the north of the Indian subcontinent as early as forty-two hundred years ago, along the Indus. But the so-called Indus civilization collapsed about four hundred years later. Eight hundred years after that, new states appeared, now along the Ganges River, too, as iron technologies allowed the clearing of forests, so agriculture expanded and populations boomed. By 500 BCE, powerful cities and states were appearing, some of which had conquered smaller city-states. Within the next two hundred years, the huge kingdom of Magadha appeared, with a capital, Pataliputra, near modern Patna. At its height, Pataliputra may have had a million people, making it as large as imperial Rome. Magadha was conquered by the Mauryan dynasty around 320 BCE, in the aftermath of Alexander the Great's unsuccessful invasion of north India in 327 BCE. It has often been claimed that Kautilya, the author of the *Arthashastra*, was the prime minister of the first Mauryan emperor, Candragupta Maurya (who ruled from 320 to 298 BCE), but the *Arthashastra* was probably written several centuries later.

The *Arthashastra* begins, like many manuals of statecraft, by arguing that the worst situation for everyone is that of statelessness, of having no ruler. A world in which no one can punish wrongdoers "gives rise to the law of the fish—for in the absence of the dispenser of punishment a weak man is devoured by a stronger man, and, protected by him, he prevails."[6] This is a convenient argument for rulers, of course, but it also captures a more general truth: even for most peasants, there were advantages to living within an orderly state.

Here is how the *Arthashastra* summarizes the main tasks of rulers:

Agriculture and animal husbandry, along with trade, constitute Economics. It is of benefit because it provides grain, livestock, money, forest produce, and labor. By means of that he [the ruler] brings under his power his own circle and his enemy's circle using the treasury and the army. What provides enterprise and security…is punishment [*danda*, or the ruler's scepter]; its administration is government. Government seeks to acquire what has not been acquired, to safeguard what has been acquired, to augment what has been safeguarded, and to bestow what has been augmented on worthy recipients. On it depends the proper operation of the world. Punishment, therefore, is the foundation of the three knowledge systems.[7]

Clearly, this is all about mobilization, about the pumping mechanisms that drive flows of energy, labor, and wealth from farmers and laborers and artisans to society's rulers in order to maintain a stable state. Much of the manual gives advice about collecting taxes, choosing officials, forming and supplying armies and prisons, and ensuring that peasants can produce enough wealth for society to flourish.

Good information was vital to mobilization. In fact, successful mobilization meant having *more* information than those from whom you were mobilizing resources. So much of the *Arthashastra* describes how to build networks of spies, keep court records, and record the government's resources and assets. Censuses were vital. The chief collector of revenue was to record the total number of villages and classify them by their wealth and the amount of grain, animals, money, forest produce, and labor they supplied, as well as the number of soldiers. City managers were advised to "find out the number of individual men and women within each [group of households] in terms of their castes, lineages, names, and occupations, as well as their income and expenditures."[8] Local tax collectors were to keep records of how many people were "farmers, cowherds, traders, artisans, laborers, and slaves." They also had to list other, smaller groups, including magicians, brothel keepers, tavern owners, soldiers, doctors, and officials. Other officials kept records of horses (listed by age, color, health, and origin), elephants, and other important resources.[9]

States, like living creatures, are complex adaptive systems, so they share many features with living organisms, and many writers have noted the similarities. In the introduction to *Leviathan,* Thomas Hobbes described the state as a huge monster or leviathan:

> An Artificiall Man, though of greater stature and strength than the Naturall...in which, the Soveraignty is an Artificiall Soul...The Magistrates, and other Officers of Judicature and Execution, artificiall Joynts; Reward and Punishment...are the Nerves...the Wealth and Riches of all the particular members, are the Strength; Salus Populi (the Peoples Safety) its Businesse; Counsellors...are the Memory; Equity and Lawes, an artificiall Reason and Will; Concord, Health; Sedition, Sicknesse; and Civill War, Death.

The main features of states do indeed parallel those of living organisms. Like the cells of living organisms, states have semipermeable borders, creating a protected internal region. Flows through the border are vital to the state's survival, so they are carefully monitored. States also have a "metabolism" that mobilizes flows of energy and resources and distributes them so as to keep the state functioning by supporting elites (the "worthy," as the *Arthashastra* puts it) and the armies and bureaucracies that defend and manage the state. For states, as for living organisms, the ultimate source of most energy flows is photosynthesis, which allows farmers to capture energy from sunlight. In states, as in living organisms, flows of energy must be managed with care. Too small, and the state starves. Too large, and subjects revolt or starve, and flows of energy and resources dry up. Just as living organisms maintain electrochemical gradients that drive flows of energy, states maintain gradients of persuasion and coercion. They use law, education, and religion to persuade their subjects that their power is just. But they also maintain armies and disciplined groups of coercers so they can compel obedience when persuasion fails. This is why the *Arthashastra* treats punishment (*danda*) as the foundation of statehood. Coercion was fundamental to mobilization in all agrarian civilizations, which helps explain the importance of warfare and the pervasiveness of physical punishments in society and within households and families.

States, like living organisms, keep track of information about their resources and enemies so they can constantly adjust to unstable environments. Staying alert to dangers and tracking flows of wealth requires some method of recording information, whether you are a bailiff, a spy, or a census taker. That is why all states have evolved some form of writing, even the Inca empire in South America, whose writing took the form of knotted ropes, or *qipu*. Everywhere, writing evolved as a way of recording politically useful information. States have rules, just as cells have genomes. In states, the rules can be found in law books, in the pronouncements of rulers and local officials, in manuals such as the *Arthashastra,* carved on stone pillars, in the collective wisdom of rulers and officials, and embedded in religious traditions.

If we think of states as a sort of genus or type of political organism, we can also argue that traditional states evolved over time, as rulers and officials learned new methods of statecraft and acquired new political, military, and bureaucratic technologies. Indeed, the history of states and agrarian civilizations over several millennia has its parallels with the history of the biosphere, as states entered new niches and evolved new methods of rule and new political technologies, as some states vanished, as new genera of states evolved, and as some states got larger and larger and acquired increasing power and knowledge.

The Spread of Agrarian States

States, like farming, appeared independently in different parts of the world. Not surprisingly, they appeared where farming had already flourished for centuries or millennia and was sufficiently developed to support large populations, large surpluses, networks of commerce and trade, and towns and cities. But states, and all the bits and pieces that go along with them, did not appear in all farming regions. In some, such as Papua New Guinea or along the Mississippi River, farming generated large villages and modest forms of power but was not productive enough to support large cities or states.

As with farming, we can trace the spread of agrarian civilizations within the different world zones almost as if we were watching the spread of an infectious disease.

Five thousand years ago, states could be found only in southern Mesopotamia and along the Nile. But already, they were diversifying. In Mesopotamia, the earliest states were based on single cities that seemed constantly at war. Along the Nile, the first states seem to have been larger, and the cities less important. Within the next thousand years, as populations grew and statecraft evolved, the states of southern Mesopotamia became more powerful and controlled larger areas. By four thousand years ago, there were states south of Egypt along the Nile Valley, in Sudan, as well as along the Indus Valley, in the north of the Indian subcontinent, in Central Asia, and in northern China, along the Huang He or Yellow River. A thousand years later, by 1000 BCE, states could be found around much of the eastern

Mediterranean; in southern China, particularly along the Yangtze River; and in parts of Southeast Asia. Powerful chiefdoms that would eventually evolve into full-blown state systems could also be found in Europe and in West Africa. Two thousand years ago, there were also states and agrarian civilizations in the American world zone, particularly in Mesoamerica and the Andes, and they had the same basic metabolic machinery as the states of Afro-Eurasia.

States and empires were becoming more powerful and wealthier. But they were also reaching over larger areas and controlling larger and more diverse populations as the technologies of government evolved. The Estonian scholar Rein Taagepera has tried to calculate the increase in the areas under states. By his estimates, the earliest states covered a tiny part of the Earth in 3000 BCE, perhaps just one-tenth of a megameter. (A megameter is equal to one million square kilometers, or about the size of the modern state of Egypt.) Between 2000 and 1000 BCE, the area under states increased to perhaps one or one and a half megameters, but this was still the equivalent of only about 1 percent of the area ruled by states today. Most of the world was still inhabited by independent farming villages and foragers.

The millennium between four thousand and three thousand years ago (between 2000 and 1000 BCE) reminds us that states could fall as well as rise. In the Indus Valley, in modern Pakistan, an entire system of states broke down, leaving behind only rich archaeological remains and tantalizing inscriptions that have not yet been deciphered. But after 1000 BCE, the momentum returned and new states appeared in new regions, while older state systems flourished and expanded. The Achaemenid Empire, founded by the Persian emperor Cyrus around 560 BCE on the remains of the Assyrian Empire in northern Mesopotamia, probably counts as the first mega-empire. At its height, it may have controlled six megameters. Two centuries later, the Mauryan Empire in northern India may have extended over three megameters, while in China, the Han Empire was as large as the Achaemenid Empire. By two thousand years ago, when the Roman and Han Empires flourished, the first state systems were appearing in Mesoamerica and the Andes, though they were smaller and less populous than the mega-empires of the Afro-Eurasian world zone. Taagepera estimates that two thousand years ago, state systems controlled about sixteen megameters, or about 13 percent of the land surface of the Earth.

The spread of states and civilizations stimulated new forms of collective learning as technologies, commodities, ideas, religions, and philosophies diffused over vast areas within the larger world zones. The expansion of populations, trading systems, and state systems was driven not just by the increased flows of food and energy from farming but also by innovation. With more people living in a greater diversity of environments, information and innovations accumulated faster than ever before. Particularly important were technologies that accelerated exchanges, such as new forms of money or improved ships or roads. The empires of Afro-Eurasia were all great road-builders. Roads were, after all, the arteries of empires. Rulers built roads so their armies and merchants could move faster and farther, but they also established courier systems so that they could learn quickly of revolts or enemy threats. The Royal Road from Susa in Persia to Sardis near modern Ephesus was built by the Achaemenid emperor Darius and described by Herodotus. It extended over twenty-seven hundred kilometers and allowed couriers using relays of fresh horses to cover in seven days a distance that would take walkers ninety days.

Writing allowed rulers to store important information about their empires and their subjects. New military technologies, such as better horse harnesses or camel saddles or more powerful catapults or faster chariots, transformed warfare, while improved communications by land and sea transformed commerce and eased the transport of farm produce. From the time of ancient Sumer, new metallurgical technologies spread through-out the Afro-Eurasian world zone, beginning with bronze, an alloy of copper and tin. From about three thousand years ago, furnaces were efficient enough to smelt iron, which was tougher than bronze and also cheaper, because iron ores were much more common and accessible than tin or copper ores. In the Iron Age, from 1000 BCE, metals were used for weapons, farming implements, harnesses, carts, and carriages, and even for ordinary household goods such as pots and pans.

Collective learning shaped educational, philosophical, and scientific thought, and lay behind the rich theologies of the major state religions, all of which incorporated origin stories in their accounts of the world. Most states tried to influence the religious ideas of their subjects, so they built temples and sup-ported official priests. Often, they persecuted shamans or other religious figures who preserved unofficial religious beliefs and practices. The earliest states worshipped local deities, but as states expanded over larger areas, their gods, too, seemed to acquire greater powers and greater reach. In the largest empires, we see the emergence of supreme deities such as the Zoroastrian god Ahura Mazda, the supreme god of the Achaemenid Empire. These were gods whose worshippers saw them as universal rul-ers, just as the empires that worshipped them claimed to rule over the known world. All the major world religions, including Judaism, Christianity, and Islam, as well as the religious tradi-tions of Rome and Greece, of Hinduism, Buddhism, and Confu-cianism, and the religious traditions of American empires, incorporated superhuman gods. And for the most part, rulers and the leaders of institutionalized religious traditions worked closely together because they understood how powerful reli-gious beliefs could be as a way of generating support for systems from which they both benefited.

Skillful rulers learned many ways of increasing their wealth. They tried to protect peasants from overexploitation, because they understood that most of their wealth came from peasant villages. It was dangerous to oppress peasants too much and sensible to protect them from enemy armies or predatory landlords and support them with grain stores when crops failed. As the *Arthashastra* pointed out, peasants were the economic foundation of each state, so wise rulers wanted peasants to prosper. Skillful rulers also encouraged international trade in order to get rare and valuable strategic goods such as precious jewels or silks for the wealthy, tin to make bronze, or even grains to feed their cities. Many also traded people, as the capture and sale of slaves as laborers, servants, and soldiers flourished in the steppes and the huge slave markets of the eastern Mediterranean and Central Asia. Those rulers who profited most from trade invested in markets and caravansaries, protected merchants, and built roads, waterways, and harbors to move goods faster and farther.

As states expanded, so, too, did networks of exchange. By four thousand years ago, Mesopotamian cities were already trading with India, Egypt, and Central Asia, while parts of Central Asia were trading with China. By two thousand years ago, such networks carried large amounts of goods, including silks, coins, glassware, and spices, right across Afro-Eurasia over the land routes known as the Silk Roads and through the sea routes of the Indian Ocean. These international exchange networks also carried goods no one wanted, including diseases such as smallpox and the bubonic plague. Plagues, such as those under the Byzantine emperor Justinian I, about fifteen hundred years ago, may explain the slowing of population growth between two thousand and one thousand years ago in the more densely settled regions of Afro-Eurasia.

By two thousand years ago, there were large empires right across Afro-Eurasia. They included the Roman, Sassanian, Kushan, Mauryan, and Han Empires. And there were many smaller semidependent states in between. During the next millennium, between two thousand and one thousand years ago, some of the larger empires collapsed, including the largest of all, the Roman and Han Empires. Disease and imperial break-down slowed growth for almost a millennium. But by a thousand years ago, there were new signs of growth. Villages, cities, and trade networks expanded in previously underpopulated regions of South China, northern Europe, and Africa. Perhaps most astonishing of all was the rise of new political systems associated with a new world religion, Islam, in the eighth century CE.

Four centuries later, early in the thirteenth century CE, the Mongol Empire was created by pastoral nomads led by Genghis Khan. Though it lasted for less than a century, it was the largest empire that had existed so far and the first to reach across the whole of Afro-Eurasia, from Korea to Eastern Europe. In the Americas, the first true state systems appeared about two thousand years ago, in Mesoamerica and in the Andes. Many American states, like those of the Maya, were based on single cities, like the city-states of Sumer three thousand years earlier. By the time of the Mongol Empire, there were also imperial systems in the Americas that controlled many cities and large territories. They included the predecessors of the Aztec and Inca Empires.

Measuring Change in the Agrarian Era of Human History

In the agrarian era, for the first time, we have just enough information to try to measure some fundamental changes in human history. We can try to estimate how human societies used energy and how energy and increasing complexity were linked in human history, as they were in the histories of stars and the biosphere. The figures in the appendix offer some very rough measures of the role of energy in human history and its impact on human lives. The figures are very tentative, of course, but they are based on some of the most careful estimates we have of large-scale changes in human history. And the story they tell is important and can help us see the broader shape of human history.

In the previous chapter, we saw that human populations increased during the Paleolithic, but very slowly, perhaps by a quarter of a million people every millennium in the last twenty thousand years of the most recent ice age. The figures in column B of the appendix show the sharp acceleration in population growth after the introduction of agriculture. Between ten thousand and five thousand years ago, human populations quadrupled, and then, between five thousand and two thousand years ago, they multiplied again by ten times. So, over the entire period from ten thousand to two thousand years ago, human populations increased by about forty times, at an average rate of twenty-five million every millennium, or about one hundred times the average rate of growth of the Late Paleolithic.

Such rapid population growth was made possible by huge increases in the energy consumed by our species (column C). By two thousand years ago, humans were using seventy times the amount of energy they consumed at the end of the last ice age. This colossal energy bonanza from farming paid for population growth, for entropy's various complexity taxes, and finally for the wealth of the rich and powerful. There is little sign that it improved the lives of most humans.

Most of the energy bonanza paid for growing populations. But not all, because, as column D shows, there was a slight increase in the amount of energy consumed per person after five thousand years ago. We can't measure precisely how that extra energy was allocated, but what we already know of the evolution of agrarian societies suggests the most important ways it was used. It went, first, to pay for increasing complexity. Column F in the statistical appendix offers a very rough measure of increasing complexity, on the assumption that the size of the largest cities indicates the extent of the human capacity to build, maintain, and pay for complex social and technological structures. After all, cities, like civilizations in general, depend on a huge amount of organization and massive expenditures on buildings, roads and highways, irrigation canals, palaces and temples, officials, police, markets, and soldiers. We can regard these expenses as part of the complexity taxes paid to entropy. There was also a sort of garbage tax paid to entropy. This was energy from which no one really benefited, and it includes the energy wasted during wars and natural or epidemiological disasters.

We know that some of the extra energy from farming also went into improving the lives of the elite groups who made up something like 10 percent of most agrarian civilizations. Elites controlled great wealth, and it is likely that even the slow rise in life expectancy (column E) was largely confined to the powerful and wealthy. So at least some of the energy bonanza from agriculture helped improved some human lives. But after all these other outlays, there was little left to raise the living standards of the rest of the population. That is why all the evidence we have suggests that, though people surely enjoyed occasional luxuries, most of the time, most of them lived close to subsistence level throughout the agrarian era. The French economist Thomas Piketty has estimated that in most European countries as late as 1900, 1 percent of the population owned about 50 percent of national wealth, and 10 percent of the population accounted for 90 percent of national wealth. The other 90 percent of the population made do with just 10 percent of national wealth. There was really no middle class in the modern sense because "the middle 40 percent of the wealth distribution were almost as poor as the bottom 50 percent. The vast majority of people owned virtually nothing, while the lion's share of society's assets belonged to a minority."[10]

If this distribution of wealth was typical of most agrarian civilizations, it supports the general conclusion that the energy bonanza from farming improved the lives of no more than a tenth of all human beings. But that's surely the story of most gold rushes. To spread wealth more widely would take one more energy bonanza, one even more spectacular than the energy bonanza from farming. The next chapter describes the changes that prepared the way for threshold 8, the threshold that would lay the foundations for the astonishing energy-rich world of today.

CHAPTER 10

On the Verge of Today's World

*The discovery of America, and that of a passage to the
East Indies by the Cape of Good Hope, are the two greatest and most important events recorded in the history of
mankind. . . . By uniting in some measure the most distant parts of the world, by enabling them to relieve one
another's wants, to increase one another's enjoyments,
and to encourage one another's industry, their general
tendency would seem to be beneficial. To the natives,
however, both of the East and West Indies, all the commercial benefits which can have resulted from those
events have been sunk and lost in the dreadful misfor-
tunes which they have occasioned.*

— ADAM SMITH, *AN INQUIRY INTO THE NATURE AND
CAUSES OF THE WEALTH OF NATIONS*

*I sell here, sir, what all the world desires to have—
POWER.*

— MATTHEW BOULTON, THE MAJOR INVESTOR IN
JAMES WATT'S IMPROVED STEAM ENGINE

When describing previous thresholds of increasing complexity,
we have offered some educated guesses about the Goldilocks
conditions that made them possible. As we approach today's
world, we can see with much more precision how new Goldilocks

conditions accumulated, eventually preparing the way for the astonishing burst of innovation that would create today's world, the world of the Anthropocene.

The World Six Hundred Years Ago

By 1400 CE, human populations had grown from about five million at the end of the last ice age to one hundred times as many, or almost five hundred million people. There were still large regions, in Australasia, parts of Africa, central Eurasia and Siberia, and the Americas, where populations were small, and most people lived from foraging or hunting or herding or pastoral nomadism. But most humans now lived within agrarian civilizations and depended directly or indirectly on farming. Indeed, most humans *were* farmers. Many parts of the world were filling up with farmers, just as, ten thousand years earlier, some regions had filled up with foragers. Even the Pacific was filling up, as Polynesian sailors set off on the dangerous migrations that took them to most parts of the Pacific. Aotearoa (New Zealand), the last large area of farmable land in the Pacific, was settled about seven hundred years ago.

As human numbers increased, so did the pressure to find new land, new resources, new sources of wealth. Siberian foragers and reindeer herders came under growing pressure from tax officials, fur traders, merchants, and pastoral nomads to trap and sell furs and walrus tusks and forest commodities. In Australia, where there were no agrarian states to press for more resources, population growth forced people to increase production. In fertile regions, such as around modern Sydney, tribal territories shrank as populations grew, and local communities had to develop more specialized and intensive technologies. In Sydney's harbor, in recent centuries, women fished with lines made from kurrajong bark and special hooks carved from "turban" shells that let them take fish from deeper waters. They fished at night from bark canoes known as *nowie*, in which they lit fires to keep themselves and the babies at their breasts warm. In 1770, Joseph Banks, who sailed with Captain Cook, saw Sydney's Botany Bay full of the twinkling lights from *nowies*.[1] In some regions of Australia, there were semipermanent villages and the beginnings of farming.

In some of the larger Pacific islands, such as Hawaii, Tonga, and New Zealand, farming was productive enough to support small townships and small states. In Central America and in the Andes, agriculture had spread over large enough regions to support not just large states but the first American imperial systems. The core region of the Aztec Empire, which evolved rapidly in the fifteenth century, was in modern Mexico. Its capital, Tenochtitlán, was where Mexico City is today. The heartlands of the Inca Empire, its contemporary, were on the slopes of the Andes, in Ecuador and Peru. The Inca capital, Cuzco, was in the southeast of modern Peru.

Population pressure and competition to mobilize new resources were felt most acutely in Afro-Eurasia, the oldest, largest, most populous, and most diverse of the world zones. As they looked for more energy and resources, rulers, entrepreneurs, and land-hungry peasants competed for new farmable lands and new forms of wealth, including furs, spices, and minerals.[2] And they were always willing to push aside foragers if necessary. These pressures drove peasants to settle in lands they might once have scorned, in the north of Scandinavia, for example, or in parts of Ukraine and Russia on the edges of the arid Eurasian steppes. Mobilizational pressure thickened and diversified networks within Afro-Eurasia, increasing their size and the wealth and diversity of the goods and ideas they exchanged through the Silk Roads or through the maritime routes of the Indian Ocean.

In 1400 a concentrated band of people, cities, and farmlands stretched from the Atlantic Ocean, along both sides of the Mediterranean, through Persia and parts of Central Asia, and into India, Southeast Asia, and China. The richest and most populous empire in 1500 was ruled by the Ming dynasty in China. In the early fifteenth century, the Ming emperor Yongle sent out vast fleets, captained by a Muslim eunuch, Zheng He, to travel through the Indian Ocean to India, Persia, and the rich ports of East Africa. Zheng He's ships were some of the largest and most sophisticated that had ever been built, and their many voyages provide an interesting foretaste of the globalization that was just around the corner. But after 1433, under a new emperor, Hongxi, the Ming abandoned these expeditions. China was wealthy and pretty self-sufficient, so Zheng He's expeditions had little commercial value. Besides, they were extremely expensive. The new emperor and his advisers decided that the money spent on them could be put to better uses, such as defending the empire's northern borders from pastoral nomadic invaders.

Rulers with fewer resources and smaller populations had more reason to seek wealth beyond their borders. Expanding particularly rapidly in the fifteenth and sixteenth centuries was the young empire of Muscovy. Its governments built lines of forts that pushed its frontiers south toward the fertile but arid grasslands north of the Black Sea, southeast toward the Silk Road markets of Central Asia, and east into the rich fur and mineral quarries of Siberia. The Ottoman Empire was the most powerful empire in the Muslim world. By the sixteenth century, its power reached into southeastern Europe, through Mesopotamia, and across northern Africa. After the conquest of Egypt in 1517, it also controlled the lucrative trades from the Indian Ocean into the Mediterranean and on to Europe. In the same century, a rival Muslim empire emerged in the Indian subcontinent: the Mughal Empire, founded by Babur, a descendant of the Mongolian emperor, Genghis Khan. In Africa, there were powerful states and empires north of the Sahara, along the Nile, and in West Africa, as well as along the eastern coast, which was dotted with wealthy trading cities. Europe lay at the western edge of the Eurasian landmass, far from the rich streams of commercial wealth that passed through the Mediterranean and Indian Oceans. The Venetians managed to tap those trade flows, but it was not easy. In 1500, Europe's most powerful empire was the Holy Roman Empire, a ramshackle collection of states, bishoprics, and principalities linked by marriage and conquest and extending from Austria, through Germany, and into the Netherlands and Spain.

In 1400 the world was still divided into distinct world zones between which there were no significant contacts. But growing populations and growing mobilizational pressure ensured that, sooner or later, the ocean membranes between the world zones would be breached. Who would do this and when remained uncertain, though the intense mobilizational pressures in the Afro-Eurasian zone made it extremely likely that the breach would come from within this zone.

In 1492, the ocean between the two largest world zones was finally crossed by an expedition led by a Genoese navigator, Christopher Columbus. Columbus had persuaded the rulers of Spain to back his hunch that there was a quick route across the Atlantic from Europe to the rich markets of eastern Asia. Over the next three centuries, the membranes separating Australasia and the Pacific zone would also be breached, and for the first time in human history, people would start exchanging information and ideas, goods, people, technologies, religions, and even diseases across the entire world.

The change was transformative. For the first time since plate tectonics had created the single supercontinent of Pangaea, two hundred and fifty million years ago, genes, organisms, information, and diseases could flow within a single worldwide system. The world historian Alfred Crosby described this ecological revolution as the "Columbian Exchange," and he showed that globalization would transform the biosphere as much as it transformed human history.[3] In *The Communist Manifesto,* Marx and Engels argued that these changes kick-started modern capitalism.

> The discovery of America, the rounding of the Cape, opened up fresh ground for the rising bourgeoisie. The East-Indian and Chinese markets, the colonization of America, trade with the colonies, the increase in the means of exchange and in commodities generally, gave to commerce, to navigation, to industry, an impulse never before known, and thereby, to the revolutionary element in the tottering feudal society, a rapid development.

So powerful was the shock from linking the different world zones that, within just a few centuries, human societies had crossed the eighth threshold of increasing complexity. The change was fast because it occurred in a globalized world. In the past, collective learning had worked at local or regional scales, which is why it took ten thousand years for farmers to spread around the planet. In a world of global networks, it took just a few centuries to transform much of Earth. This was a change as momentous as anything that had happened in the entire four-billion-year history of the biosphere. Suddenly, humans found themselves linked within a single global sphere of thought: the noösphere. By the twentieth century, the noösphere had become a disruptive force for change within the entire biosphere.

Creating a Single World System

European navigators were the first to link up the major world zones. That simple fact gave European rulers and entrepreneurs a colossal advantage for several centuries, because Europe, which had once been far from the great hubs of wealth and power, now controlled the gateways through which passed the largest flows of wealth and information in human history.

European navigators broke through to the other world zones because they did *not* enjoy easy access to the rich markets of South and Southeast Asia. That meant they had to take risks if they were to get their share. Most important, they would have to bypass the Ottoman traders who dominated the Mediterranean. That is one reason why, in the mid-fifteenth century, Portuguese governments began to send highly maneuverable caravels armed with cannons to probe around the western coast of Africa. The caravels, with their lateen sails inspired by Islamic models and their compasses and cannons derived from Chinese inventions, were themselves examples of the intellectual synergies accumulating within the Afro-Eurasian world zone. By the 1450s, Portuguese navigators had already established profitable maritime trades with the Mali Empire for the gold, cotton, ivory, and slaves that had previously been moved by camel caravan across the land routes of the Sahara.

These modest successes encouraged rivals. The Genoese navigator Christopher Columbus was one of them. Columbus had persuaded the Spanish rulers Ferdinand and Isabella to back him in seeking a more direct westward route to Asia by sailing far out into the Atlantic. He believed, incorrectly, that the distance to China across the Atlantic was much less than many had supposed. Ferdinand and Isabella gambled on his idea because they knew that if Columbus was right, the rewards would be stupendous. On October 12, 1492, his ships reached an island he called San Salvador, in the Bahamas. To the end of his life, he was sure he had arrived in Asia, or the Indies, and that is why he described the people he met as Indians. That is also why he was puzzled by their nakedness and apparent poverty[4] and the absence of kimonos and silken gowns. Captives led him to Cuba, where he found small amounts of gold, and that was just enough to persuade Ferdinand and Isabella to fund more voyages. Columbus's voyages brought the American and Afro-Eurasian world zones into regular contact for the first time. In 1498, just six years after Columbus's first transatlantic voyage, a Portuguese captain, Vasco da Gama, showed that it was also possible to reach Southeast Asia by sailing around the southern tip of Africa. The Indian Ocean was not a vast, enclosed lake, as many had supposed.

Many, perhaps most, of the early encounters between people from the different world zones were violent, chaotic, and destructive. Suspicion of strangers played a role. But so did the many differences in population densities, technologies, patterns of social and military organization, and even resistance to diseases that had accumulated over many millennia. There were winners and losers, and for the losers, the outcomes could be catastrophic. Like the appearance of the first oxygen atmosphere or the sudden death of the dinosaurs, this was an example of what the Austrian economist Joseph Schumpeter termed *creative destruction*—the constant, often violent replacement of the old by the new, which Schumpeter saw as the very heart of modern capitalism. Many societies were ruined, and many lives destroyed. But there was creation, too, because the sheer scale of the first global-exchange networks synergized collective learning on a planetary scale, releasing huge flows of information, energy, wealth, and power that would eventually transform human societies throughout the world.

Almost all the advantages lay with the resource-hungry states and empires at the western edge of Afro-Eurasia whose ships had first broken through the barriers between the world zones. They exploited those advantages with predatory glee and ruthless efficiency. Within fifty years of Columbus's first voyage, the Portuguese had used their armed caravels to build fortified strong points that bolted together a trading empire in the

Indian Ocean. The risks for merchants and sailors were huge, but so were the potential profits. In the Americas, Spanish conquistadores, such as Hernán Cortés and Francisco Pizarro, seized control of the rich civilizations of the Aztecs and Incas. They did so with tiny armies that exploited political divisions within both empires. But they were assisted by the devastating impact of European diseases such as smallpox that may have killed up to 80 percent of the population in America's major empires and ruined ancient social structures and traditions. At huge cost to others, the conquistadores really had struck gold, and they enriched themselves and their home societies.

In the Americas, Spanish conquerors found more than gold and silver. They also found land that could be used to grow crops such as sugar, for which the European appetite was huge and growing. Spaniards (including Columbus's own relatives) had already shown how to produce sugar cheaply in the Canary Islands, where it was grown on plantations worked by slaves. These plantations gave a foretaste of the profits that would be made in the Americas, often using violence of the most brutal kind.

In the 1540s, at Potosí in modern Bolivia, Spanish merchants found a mountain of silver. At first they exploited it using traditional systems of forced labor inherited from the Inca. But death rates were so high that they soon began to use imported African slaves. Mule trains carried silver to the Mexican port of Acapulco, where it was minted into silver pesos, the world's first global currency. Many pesos flowed across the Atlantic to Europe, where they buoyed local economies, as the Spanish government used them to pay off debts to Dutch or German creditors. Pesos also traveled across the Pacific in the Manila galleons to the Spanish-controlled city of Manila. Here, Spanish merchants and officials traded them for Chinese silks, porcelains, and other goods supplied by Chinese merchants, which they resold in the Americas and Europe at huge profits. This was classic arbitrage trading. Merchants bought where goods were cheapest and sold where they were dearest, and they profited enormously because the gap between production costs and sale prices could be extremely wide in the world's first global markets. The booming Chinese economy needed silver and valued it highly, so silver was worth twice as much in China as it was in Europe, and slave labor in the Americas kept its production costs low. High-quality silk, by contrast, was commonplace in China but rare and immensely valuable in Europe.

As long as their ships avoided shipwrecks and pirates, European merchants and their backers could make huge profits by exploiting the steep price gradients on the first global-exchange networks. What the Portuguese and Spanish had started, the Dutch and English continued in the seventeenth century as they seized Portuguese forts in Asia and began to nibble away at Spanish and Portuguese colonies in the Caribbean and North America.

Information flowed down these gradients alongside wealth, and information would prove equally important. The invention of efficient new ways of printing by Johannes Gutenberg in the mid-fifteenth century magnified the impact of new information flows. Almost thirteen million books were published between 1450 and 1500, and more than three hundred million between 1700 and 1750.[5] Books, and the information they stored, ceased to be a rare, pricey luxury and became an everyday acquisition for people with education. And, just as arbitrage profits stimulated European commerce, huge new flows of information stimulated European science and technology.

European navigators found new continents and islands, saw new constellations in the southern skies, and encountered peoples, religions, states, plants, and animals never mentioned in ancient texts. The tsunami of new information shook up education, science, and even religion throughout Europe, because this was the region through which new information flowed first and fastest. That information forced European scholars to question ancient science, and even the Bible. It began to undermine traditional origin stories. In sixteenth-century England, Francis Bacon argued that science and philosophy should no longer rely mainly on ancient texts but should actively seek out new knowledge, like Europe's navigators: "By the distant voyages and travels which have become frequent in our times many things in nature have been laid open and discovered which may let in new light upon philosophy."[6] "There is," wrote Joseph Glanvill in 1661, "an *America* of secrets, and [an] unknown *Peru* of Nature" waiting to be found.[7]

As David Wootton, a modern historian of the scientific revolution, puts it, "the idea of discovery is...a precondition for the invention of science."[8] Study the world itself rather than what has been *said* about the world. Learn how to "conquer Nature by obeying her," as Bacon put it. This was very much in the manipulative spirit of modern science and technology. In the seventeenth century, many scholars began to understand that they were living through an intellectual as well as a geographical and commercial revolution and that new knowledge was increasing human power over the natural world. "As to our work, we are all well agreed," wrote a member of the Royal Society in 1674, "...that it is not to whiten the walls of an old house, but to build a new one."[9] In the eighteenth century, European thinkers of the Enlightenment era began to see purpose, meaning, and "progress" in new knowledge. The idea that humans should transform and "improve" the world began to shape science, ethics, economics, philosophy, commerce, and politics.

The world of thought was transformed. David Wootton describes the change vividly. In Shakespeare's time, even the most educated Europeans generally believed in magic and witchcraft, in werewolves and unicorns; they believed that Earth stood still and the heavens turned around it; that comets portended evil; that the shape of a plant advertised its medicinal powers because God had designed it to be interpretable; that *The Odyssey* was a true history.[10] A century and a half later, when Voltaire was alive, educated Europeans thought very differently. Many collected or read about experimental instruments such as telescopes, microscopes, and air pumps; they thought of Newton as the greatest of scientists; they knew Earth orbited the sun; they did not take magic, the histories recounted in ancient legends, the stories of unicorns, or (most) stories of miracles seriously; they believed in the advancement of knowledge and something like progress.

New information provided the intellectual bricks and mortar for new types of knowledge. As Isaac Newton developed his laws of gravity, he had access to an unprecedented range of information. He could, for example, compare how pendulums swung in Paris with how they swung in the Americas and Africa. No previous generation of scientists could have tested their ideas so thoroughly or within such wide and varied networks of information.

> Newton's achievement can be tied to the vast increase in general knowledge that overseas trade and exploration had brought to Europeans. The courage to generalize, to arrive at universals about the natural world, owes much to the immense quantity of information—and self-confidence—that European mastery of the great seas gave land-bound thinkers like Isaac Newton.[11]

Dazzling new flows of wealth and information had one more powerful effect: they stimulated the commercial forms of mobilization often described as *capitalism* that were driven by gradients of both wealth and information. For the most part, traditional rulers had mobilized wealth with the threat of coercion, the promise of protection, and appeals to religious and legal authorities. But in all civilizations, merchants had also mobilized a lot of wealth through commerce. Commercial mobilization depended on arbitrage, on buying cheap in one region and selling dear somewhere else. To succeed, merchants needed wealth to invest and information about what to invest it in. The steep gradients of wealth and information in the first global-exchange networks opened such vast commercial opportunities for European merchants and entrepreneurs that their wealth and political influence increased until even emperors, such as the Holy Roman emperor Charles V, began to borrow from merchants.

European rulers were generally keener to work with merchants than traditional rulers such as the Ming emperors of China had been because most European states had modest resources, fought endless wars, and were constantly short of cash. And rulers who borrowed from merchants were naturally eager to support commerce. In this way, there emerged a close symbiotic relationship between European traders and rulers. Rulers protected and supported commerce, and in return they got the right to tax and profit from commercial wealth. This was the earliest and crudest form of capitalism, a system admired by European economists from Adam Smith to Karl Marx.

The emerging partnership between European governments and entrepreneurs took many forms. The Russian trade in vodka is a case in point.[12] Distilling appeared in sixteenth-century Russia. Almost immediately, officials in the government of Ivan the Terrible (whose nickname refers to his brutal treatment of his own nobles) realized that if they could stop peasants from distilling at home (which wasn't hard to do, because distilling required a great deal of skill and equipment), they could make a lot of money, as liquor would be one of the few goods that peasants had to buy from others. It was a powerful mind-altering substance and rapidly became obligatory for peasants, who used it to celebrate the great religious and family festivals as well as marriages and funerals. But taking vodka into thousands of villages scattered over a large area was a demanding task, and one best suited to merchants. So, in partnership with its merchants, the Russian government built up a trade in vodka so profitable that by the nineteenth century, it was paying for much of the cost of the Russian army, then one of the largest in the world. Russian governments and society paid a significant entropy tax for the complex revenue pumps of the vodka trade, which eventually led to high and dangerous levels of alcoholism.

Though capitalism generated new forms of inequality, economists admired it because it was also good at generating both wealth and innovation. Many early economists understood perfectly well that the wealth traded by and generated by capitalists really represented control over compressed sunlight, over energy flows through the biosphere. That is why so many subscribed to a labor theory of value; labor, after all, was energy. But they also understood that capitalism was particularly good at encouraging innovation in control over energy. This was because merchants, unlike traditional rulers, could rarely use naked force to mobilize wealth (though they were happy to do so if they got the chance). For the most part, merchants had to use guile rather than force. That meant seeking out new information. They had to find new commodities and markets, and they had to trade efficiently and cut costs. Above all, they had to innovate if they wanted to outsmart their rivals. They had to find new ways of mobilizing and controlling flows of energy and resources. This helps explain why the increasingly capitalistic societies of Europe became both wealthier and more innovative in the centuries after Columbus first crossed the Atlantic.

Some governments, such as those of the Netherlands or Venice, were ruled by merchants, so they took commerce very seriously indeed. The British learned much from the Dutch, and briefly, in the late seventeenth century, they even had a Dutch king, William III. British governments spent huge amounts on a navy that could protect fortified trading bases and colonies in the Caribbean, North America, and, eventually, India. With naval protection, British governments and merchants made huge profits. For example, they sold arms to African rulers in return for slaves, which they transported to the Americas under appalling conditions. The slaves were traded for sugar, tobacco, and other plantation goods, whose prices were kept low because slave labor was cheap. That meant that plantation goods could be sold cheaply and profitably on the rapidly expanding consumer markets of England and Europe. The British government, like the Dutch, became increasingly dependent on revenues from trade, including customs payments. That helps explain why, in 1694, the government established the Bank of England, to make cheap loans available to British merchants, entrepreneurs, and landlords. In the eighteenth century, cheap loans encouraged agricultural innovation and helped build canals and an extensive system of coach transport. London grew into one of the world's largest cities, and British commerce boomed.

New flows of wealth and information and new forms of scientific knowledge stimulated innovation in agriculture, mining, shipbuilding and navigation, canal construction, and many other areas. They did so particularly in Western Europe. After 1500, wealth and power began to shift fast, and the former backwaters of Europe and the Atlantic region rapidly became a new hub, the center of the first global flows of wealth, information, and power.

Fossil Fuels: A Mega-Innovation

A globalized world and an increasingly wealthy and powerful entrepreneurial class supported by local rulers stimulated commerce and innovation, particularly in the Atlantic region. But, as we have seen, some innovations are more transformative than others. Not surprisingly, given Europe's increasing wealth, entrepreneurial dynamism, and information flows, the mega-innovations that would create the modern world popped up here, rather than in the older hub regions that reached across Eurasia from the Mediterranean through the Muslim world to China.

The most important mega-innovations were usually those that released new flows of energy, such as fusion or photosynthesis. Farming counts as a mega-innovation because it let farmers tap larger shares of energy flows from recent photosynthesis. Those increasing flows drove the turbulent changes of the agrarian era. But there were limits to the energy flows from farming, because it tapped only recently captured sunlight. Burn a piece of wood, eat a carrot, or harness a horse to a plow, and you are tapping energy flows captured from sunlight in the past twelve months or at most in recent decades. By the late eighteenth century, some economists in Western Europe began to suspect that European societies were exploiting these flows to the fullest. Their calculations were simple. The energy flows that powered human societies came from croplands and woodlands, with a small bonus from wind and rain. So growth meant finding more arable land and woodland. By 1800, it seemed that most farmable land was already being farmed. Adam Smith, the founder of modern economics, argued that societies would soon be using all available energy. Then growth would stall; wages would fall, and so, too, would populations as farming societies came face to face with the limits on energy flows that all other organisms do when they have filled up their niche.[13] Some societies, such as the Netherlands and England, already seemed to be pushing at these limits. In the Netherlands, farmers had to gouge farmland from the sea, while England faced growing shortages of timber for heating, housing, and shipbuilding. By Adam Smith's time, as Alfred Crosby puts it: "Humanity had hit a ceiling in its utilization of sun energy."[14]

Pressure to find new sources of energy would eventually conjure up the mega-innovations that we describe today as the fossil-fuels revolution. These gave humans access to flows of energy much greater than those provided by farming—the energy locked up in fossil fuels, energy that had accumulated not over a few decades but since the Carboniferous period, more than 360 million years earlier. In seams of coal, oil, and gas lay several hundred million years' worth of buried sunlight in solid, liquid, and gaseous forms. To get a sense of the energies locked up in fossil fuels, imagine carrying a car full of passengers over your head and running very, very fast for several hours, then remind yourself that a few gallons of gasoline pack that much energy and more (because a lot of the energy is wasted). Like a gold strike, this energy bonanza generated frenzied and often chaotic new forms of change and created and destroyed the fortunes of individuals, countries, and entire regions. Charles Dickens, Friedrich Engels, and others saw the terrible price that many paid for these changes. But from the frenzy would emerge an entirely new world.

The transformations began with technological breakthroughs that turned the energy of coal into cheap mechanical energy that could power factories, locomotives, steamships, and turbines. Many societies already knew about coal, but it was difficult to mine and transport and dirty and smelly when burned. So most people in agrarian societies preferred to get their heat energy from wood. In some regions, though, wood was scarce. In England, as populations grew, cities expanded (particularly London), and commerce boomed, demand for energy began to outstrip supplies. England was one of the first countries in the world to feel the energy squeeze. But, unlike most countries, England had a fallback. It had large reserves of coal quite close to the surface, much of it near rivers or the coast, so it could be transported cheaply and easily by sea or canals to the major cities, including London. English manufacturers and households began switching over to coal. By the seventeenth century, English brewers, brickmakers, and bakers were using coal, and Londoners began to complain about the city's foul air. By 1700, coal was producing 50 percent of English energy. By 1750, it was supplying as much energy as four million hectares of woodlands—the equivalent of almost 15 percent of the area of England and Wales.[15] Dependence on coal encouraged those who mined, transported, and sold it to produce more coal and produce it more cheaply.

But there was a problem. As demand for coal increased, coal miners had to dig deeper mines, which soon filled up with water, so getting more coal depended on building efficient pumps to drain mines. In England the incentives to solve this technological problem were greater than anywhere, so designing cheap, efficient pumps became a major goal for entrepreneurs and inventors. The combination of new science and widespread mechanical skills provided the intellectual background needed to solve the problem. Seventeenth-century scientists had begun to understand how atmospheric pressure worked, and by the early eighteenth century, that knowledge was put to use in Newcomen steam engines to pump water from coal mines.[16] But the Newcomen steam engine was inefficient and used huge quantities of coal, so it made commercial sense only in coal mines, where coal was cheap. Investors, inventors, and engineers understood that improved pumps could earn them huge profits and revolutionize the supply of coal to English homes and industries.

James Watt, the engineer who eventually solved these technical problems, was a Scottish instrument maker, well connected to engineers, scientists, and businessmen. While on a Sunday afternoon stroll in 1765, Watt suddenly figured out that he could make the Newcomen engine more efficient by adding a second cylinder that acted as a condenser. But building the improved steam engine involved cutting-edge science and technology and the ability to design and bore precisely engineered pistons that could withstand high pressures. The task was demanding and expensive. However, Watt's main backer, Matthew Boulton, sensed an opportunity and invested heavily in Watt's research. He under-stood the huge profits that could be made from a machine that turned the energy of coal into mechanical energy at a reasonable cost. By 1769, when Watt acquired a first patent on his design, competition was so intense that after Boulton bragged about Watt's prototypes to the Russian ambassador in London, Watt got a lucrative job offer from the Russian government. Watt seriously considered taking the offer, but Boulton persuaded him to stay. By 1776, the work was done.

The James Watt steam engine gave a first taste of energy flows so vast that they would transform human societies in just two centuries. Like the activation energies that kick-start chemi-cal reactions, energy from fossil fuels provided a pulse of energy that started the technological equivalent of a global chain reac-tion. Within twenty-five years, five hundred of the new machines were at work in England, and by the 1830s, coal-fired steam engines were the main source of power in British industry. English consumption of energy soared. By 1850, England and Wales were consuming nine times as much energy as Italy, and English entrepreneurs and factories had access to prime movers of colossal power. Steam locomotives could generate two hundred thousand watts of energy (yes, James Watt gave his name to the unit), or about two hundred times the energy supplied by a two-horse plow team, one of the most important prime movers of the agrarian era. More cheap energy was available than ever before. English industry took off. Coal was generating as much energy as could have been extracted from woodlands covering 150 percent of the area of England and Wales.[17]

Early Industrialization

England was the first country to benefit from the energy bonanza of fossil fuels, and production took off. By the middle of the nineteenth century, England produced a fifth of global GDP (gross domestic product) and about half of global fossil-fuel emissions. Not surprisingly, global levels of atmospheric carbon dioxide began to rise from about the middle of the nineteenth century. And as early as 1896, the Swedish chemist Svante Arrhenius recognized both that carbon dioxide was a greenhouse gas and that it was being generated in large enough amounts to start changing global climates.

But such fears belonged to the future. (Arrhenius actually thought global warming was a positive development because it might stave off a new ice age.) Meanwhile, entrepreneurs and governments in other countries wanted a share in the bonanza of cheap energy and tried to beg, borrow, or steal the new technology. Steam engines were soon being built in Europe and in the newly independent United States. As they spread, they stimulated waves of new breakthrough technologies, such as the steam locomotive and steamship, each of which cheapened transportation and spun off related innovations, particularly in the manufacture of iron and steel for rolling stock, hulls, and tracks. Entrepreneurs, engineers, and scientists explored new ways of exploiting the cheap energy from steam engines in building and textile manufacturing.

There were many powerful feedback loops. Improved steam engines allowed access to deeper mines, which lowered the cost of extracting coal, so the amount of coal that was mined increased by fifty-five times between 1800 and 1900. Cheaper coal made steam engines more economical, while steamships and locomotives slashed the cost of transporting cattle, coal, produce, and people by land and sea, which stimulated global trade. Railways increased demand for iron and steel, and innovations in steel production made it economical for the first time to use steel in mass-produced goods such as tin cans, a new way of storing and preserving foodstuffs. There were unexpected side effects. Using steam to spin and weave textiles increased the demand for raw cotton, which stimulated cotton planting in the United States, Central Asia, and Egypt. Industrial production of textiles increased demand for subsidiary products such as artificial dyes and bleaches, which kick-started the modern chemicals industry, many of whose products came from coal.

Cheap energy encouraged experimentation and investment in many new technologies. One of the most important was electricity. In the 1820s, Michael Faraday realized that you could generate an electric current by moving a metal coil inside an electric field. Large-scale electricity generation became possible in the 1860s with the invention of generators powered by steam engines. Electricity and electric motors, like the proton pumps and ATP molecules of the earliest prokaryotes, provided efficient new ways of distributing power. Transformed into electricity, power could be sent cheaply to both factories and individual homes. Lightbulbs transformed home life and factory work by turning night into day, and cities, highways, and ports began to light up at night. Electricity also revolutionized communications. At the beginning of the nineteenth century, the fastest way of sending a message by land was still by horse courier. The telegraph, invented in 1837, allowed communication at the speed of light. By the end of the nineteenth century, telephones and radios made it possible to transmit real conversations more or less instantaneously over huge distances.

New technologies revolutionized warfare and weaponry. Railways and steamships moved armies and weapons faster than ever before. In 1866, Alfred Nobel invented dynamite, a powerful new explosive. Along with improved handguns and machine guns, explosives multiplied the killing power of each soldier. The destructive power of industrial weapons became clear during the American Civil War, the first real fossil-fuels war, and steam-powered, iron-hulled ships equipped with modern weapons transformed naval warfare, allowing Britain to conquer the navies of imperial China during the Opium Wars. In the late nineteenth century, supported by the wealth, the technologies, and the energy flows of the industrial revolution, the countries of once backward Europe began to conquer much of the world during the era of imperialism.

Multiple feedback loops, most traceable, ultimately, to new flows of cheap energy, explain the extraordinary dynamism of the industrial revolution and the rapidly increasing wealth and power of the first regions to industrialize. Cheap energy enabled and stimulated innovation and investment in country after country and in many different areas of manufacturing and industry. Eventually, cheap energy from coal would encourage innovations that mobilized new forms of fossil-fuel energy from oil.

Oil, like coal, was familiar. It was extracted wherever it seeped to the surface and used to make bitumen, medicine, even incendiary weapons.[18] In the mid-nineteenth century, oil, in the form of kerosene, began to be used for lighting as an alternative to whale oil, the price of which was rising, as whales were overhunted. But mineral oil was in limited supply. Some suspected there were large amounts deep underground that could be tapped using drilling techniques imported from China, where special drills had been designed to extract rock salt. Indeed, it was known that oil was sometimes found by those drilling for salt. The first serious attempt to drill for oil was conducted by Edwin Drake in the impoverished Pennsylvania town of Titusville, beginning in 1857. On August 27, 1859, just before funds ran out, Drake's drill team struck oil. Prospectors rushed to buy up land, and within fifteen months, there were seventy-five oil wells in and around Titusville. "They barter prices in claims and shares," wrote a visitor, "buy and sell sites, and report the depth, show, or yield of wells, etc. etc. Those who leave today tell others of the well they saw yielding 50 barrels of pure oil a day.... The story sends more back tomorrow.... Never was a hive of bees in time of swarming more astir, or making a greater buzz."[19] In 1861, drillers struck the first gusher—an oil well that pumped oil under its own pressure, even producing a fatal explosion when the natural gas pumped up with the oil was ignited. Production increased to three thousand barrels a day.

Many made fortunes from oil, but not Edwin Drake, who died in poverty in 1880 despite the fact that he had helped launch the next chapter of the fossil-fuels revolution.

CHAPTER 11

The Anthropocene: Threshold 8

We're no longer in the Holocene. We're in the Anthropocene.

— PAUL CRUTZEN, OUTBURST AT A CONFERENCE IN 2000

Man the food-gatherer reappears incongruously as information-gatherer. In this role, electronic man is no less a nomad than his paleolithic ancestors.

— MARSHALL MCLUHAN, *UNDERSTANDING MEDIA*

In the twentieth century, we humans began to transform our surroundings, our societies, and even ourselves. Without really intending to, we have introduced changes so rapid and so massive that our species has become the equivalent of a new geological force. That is why many scholars have begun to argue that planet Earth has entered a new geological age, the Anthropocene epoch, or the "era of humans." This is the first time in the four-billion-year history of the biosphere that a single biological species has become the dominant force for change. In just a century or two, building on the huge energy flows and the remarkable innovations of the fossil-fuels revolution, we humans have stumbled into the role of planetary pilots without really knowing what instruments we should be looking at, what buttons we should be pressing, or where we are trying to land. This is new territory for humans, and for the entire biosphere.

The Great Acceleration

If we stand back from the details, the Anthropocene epoch looks like a drama with three main acts so far and a lot more change still in the works.

Act 1 began in the mid-nineteenth century as fossil-fuel technologies began to transform the entire world. A few countries in the Atlantic region gained colossal wealth and power and terrifying new weapons of war. A huge gap opened between the first fossil-fuel powers and the rest of the world. That gap in power and wealth would last for more than a century and start closing only in the late twentieth century.

These differences created the lopsided imperial world of the late nineteenth and early twentieth centuries. Suddenly, countries of the Atlantic region, which had been marginal for much of the agrarian era, began to dominate, and sometimes rule, much of the world, including most of Africa and much of the territory once ruled by the great Asian empires of India and China. Outside the new Atlantic hub zone, the first impact of fossil-fuel technologies was mainly destructive because the new technologies arrived in the military baggage of foreign invaders. The *Nemesis,* the first iron-hulled steam-powered gunship, with its seventeen cannons and its ability to sail fast in shallow waters, helped England win control of China's ports during the First Opium War, from 1839 to 1842. The Chinese navy, once the greatest in the world, had no defense against such weapons.

Within decades, Europe's commercial and military power had undermined ancient states and lifeways. Textile production using spinning and weaving machines powered by steam engines ruined artisan textile producers in India, which had been the agrarian era's leading producer of cotton cloth. As Britain gained political and military control of the Indian subcontinent, it locked in these imbalances by keeping Indian textiles out of British markets. Even the building of India's major railroads benefited Britain more than India. Most of the track and rolling stock was manufactured in Britain, and the huge Indian rail network was designed primarily to move British troops quickly and cheaply, to export cheap Indian raw materials, and to import English manufactured goods. In the Americas, Africa, and Asia, growing demand for sugar, cotton, rubber, tea, and other raw materials encouraged environmentally destructive plantations, often worked by quasi–slave labor. In wars that pitted machine guns against spears and assegais, European powers carved up Africa and ruled it for the best part of a century.

Europe's economic, political, and military conquests encouraged a sense of European or Western superiority, and many Europeans began to see their conquests as part of a European or Western mission to civilize and modernize the rest of the world. To them, industrialization was a sign of progress. It was part of the transformative mission, first advocated in the Enlightenment, to "improve" the world, to make it a better, richer, and more civilized place for humans.

Act 2 of the Anthropocene was exceptionally violent. It began in the late nineteenth century and lasted until the middle of the twentieth century. During this act, the first fossil-fuel powers turned on one another. In the late nineteenth century, the United States, France, Germany, Russia, and Japan began to challenge Britain's industrial leadership. As rivalries intensified, the major powers tried to protect their markets and sources of supply and keep out competitors. International trade declined. In 1914, rivalry turned into outright war. For thirty years, destructive global wars mobilized the new technologies and the growing wealth and populations of the modern era.

Other parts of the world were sucked into these wars, and

they were fought with as much brutality in China and Japan as they were in Russia and Germany. As the red mist of war descended over Europe, Africa, Asia, and the Pacific, warring governments competed to develop more destructive weapons. Science gave the combatants terrifying new weapons, some of which tapped the energies lurking within atomic nuclei. On August 6, 1945, a US B-29 Superfortress bomber flew from the Mariana Islands in the Pacific and dropped an atomic bomb on the Japanese city of Hiroshima. It destroyed much of the city and killed eighty thousand people. (Within a year, another seventy thousand had died from injuries and radiation.) On August 9, 1945, a similar weapon was dropped on the city of Nagasaki.

Act 3 includes the second half of the twentieth century and the early twenty-first century. From the bloodbath of the world wars, the United States and the Soviet Union emerged as the first global superpowers. There were many local wars, most aimed at overthrowing European colonial rule. But there were no more major international wars during the era of the Cold War. By now, all powers understood that there would be no victors in a nuclear war. But there were some close shaves. Soon after the Cuban missile crisis of 1962, President John Kennedy admitted that the odds of an all-out nuclear war had been "between one out of three and even."[1]

The four decades after World War II witnessed the most remarkable spurt of economic growth in human history. This was the period of the Great Acceleration.

Global exchanges were renewed and intensified. In the forty years before World War I, according to one influential estimate, international trade increased in value at an average rate of about 3.4 percent a year; from 1914 to 1950, that rate fell to just 0.9 percent; then, from 1950 to 1973, it rose at about 7.9 percent a year before falling slightly to about 5.1 percent between 1973 and 1998.[2] In 1948, twenty nations signed the General Agreement on Trade and Tariffs (GATT), which lowered barriers to international trade. Wartime technologies were now put to more peaceful uses. Oil and natural gas added to the energy bonanza of the nineteenth century, and so did nuclear power, the peaceful counterpart of nuclear weapons. Productivity soared, first in the leading fossil-fuel economies and then elsewhere. Consumption soared too as output rose and producers sought new markets at home as well as abroad. In wealthier countries, this was the age of the automobile, of TV, of suburban dream houses, and, eventually, of computers, smartphones, and the Internet. A new middle class started to emerge. This was also when the industrial revolution began to spread beyond the old industrial heartlands. By the early twenty-first century, industrial technologies had transformed much of Asia, South America, and parts of Africa as completely and as fast as they had once transformed European societies. As other areas of the world industrialized, their wealth and power increased. There began to appear, once again, a world with multiple hubs of power and wealth. Within two hundred and fifty years of the first modern steam engine, fossil-fuel technologies had transformed the entire planet.

During the Great Acceleration, humans mobilized energy and resources on such an unprecedented scale that they began to transform the biosphere. That is why many scholars date the dawn of the Anthropocene epoch to the middle of the twentieth century.

Transforming the World: Technologies and Science

Innovation, propelled by cheap energy, was the main driver of change. Innovations created steeper gradients of wealth and power that encouraged competition, which drove innovation, in a powerful feedback cycle. Entrepreneurs and governments hunted down the innovations that might give them an industrial or military edge and invested in the businesses and scientists, the schools, universities, and research institutes that could generate and disseminate new technologies and skills.

The wars of the early twentieth century drove a forced march of innovation. During World War I, Germany ran short of natural fertilizers, and German scientists, led by Fritz Haber and Carl Bosch, figured out how to draw nitrogen from the air to make artificial fertilizers. Nitrogen doesn't like to react, so this was not easy. Prokaryotes had solved the problem billions of years ago, but Haber and Bosch were the first multicellular organisms to successfully fix atmospheric nitrogen. The Haber-Bosch process uses huge amounts of energy to overcome nitrogen's reluctance to combine chemically, so it was viable only in a world of fossil fuels. But artificial nitrogen-based fertilizers transformed agriculture, raised the productivity of arable land throughout the world, and made it possible to feed several billion more humans. It turned fossil-fuel energy into food.

A liquid fossil fuel, oil, was first used in the late nineteenth century as a replacement for whale oil in lighting. The first internal combustion engines, developed in the 1860s and 1870s, showed how to generate mechanical force from oil. Unlike the steam engine, whose heat source was external to the engine's moving parts, in internal combustion engines, the heat from fossil fuels drove pistons or rotors or turbine blades directly. Internal combustion engines spread rapidly in the late twentieth century, largely because of their wartime use to transport soldiers and equipment and to power the first tanks. They were also installed in the first military aircraft, which pioneered the dark art of dropping explosives from the air. Once the wars ended, manufacturers of automobiles and planes turned to civilian markets to create a world in which more and more individuals owned and used cars or flew in planes. Global trade was transformed by oil tankers, container ships, and large planes.

Information lies at the heart of Anthropocene technologies.

Information technologies were transformed when governments invested in a massive expansion of education and research, and businesses and corporations funded research to develop and disseminate new products and services. To break enemy codes, wartime governments funded research into the mathematics of information and computing. This research, combined with the invention of the transistor in the late 1940s, laid the foundations for the computerization of science, business, government, finance, and everyday life in the second half of the century. Rocketry, also developed during the wars, would eventually send humans into space. Wartime governments had launched huge research programs to develop nuclear weapons. The American government's Manhattan Project developed the first atomic bombs, including the weapons dropped on Hiroshima and Nagasaki in 1945. These unleashed the energies of disintegrating uranium nuclei. The Soviet Union soon developed its own atomic weapons, helped by information leaked by spies from the Manhattan Project. Within a decade, the United States and the Soviet Union had also built hydrogen bombs, which released the much greater energies generated by proton fusion, the same mechanism that powers all stars. The first H-bomb was tested in 1952.

Much of this innovation was inspired by breakthroughs in the supercharged collective-learning environment of modern science. Albert Einstein developed his theory of relativity in the first two decades of the twentieth century. It improved on Newton's understanding of the universe by showing that matter and energy warped space and time, and this warping was the real source of gravity. Einstein also showed that matter could be converted into energy, and that insight provided the scientific foundations for nuclear weapons and nuclear power. Quantum physics, developed in the same era, gave deeper insight into the strange, probabilistic world of atomic nuclei. Without that understanding, nuclear weapons, transistors, global-positioning systems, and modern computers would not exist today. In the

1920s, astronomers such as Edwin Hubble found the first evidence that our universe began in a big bang. In biology, Darwin's idea of natural selection was combined with Mendel's understanding of heredity and the improved statistical methods of R. A. Fisher to lay the foundations for modern genetics.

These and many other new insights and technologies powered innovation and growth during the Great Acceleration. Increased productivity allowed human populations to grow faster than ever before. In 1800, there were nine hundred million humans on Earth. By 1900, there were one and a half billion. By 1950, when I was a child, there were two and a half billion humans, despite the huge casualties of the world wars. During my lifetime, human numbers have increased by another five billion. Such enormous numbers can numb the brain, so it's worth taking the time to grasp what they mean. In the two hundred years since 1800, the number of humans increased by more than six billion. Each additional human had to be fed, clothed, housed, and employed, and most had to be educated. The challenge of producing enough resources in just two hundred years to support an extra six billion humans was colossal.

Remarkably, the challenge was met, with modern technologies, modern fossil fuels, and modern managerial skills. Productivity soared in agriculture, manufacturing, and transportation. Though food and other supplies did not always get to those who needed them, enough food was produced to feed more than seven billion people. The crucial changes were in the production of artificial fertilizers and pesticides, the use of fossil-fueled farm machinery, the building of thousands of irrigation dams, and the production of new, genetically modified crops. Modern farming technologies brought new land into cultivation, increasing the farmed area from half a billion hectares in 1860 to almost three times as much in 1960.[3] Fishing trawlers equipped with powerful diesel engines, sonar detection equipment, and massive nets sucked up most of the organisms in the areas they fished. The fish catch rose from nineteen million tons to ninety-four million tons between 1950 and 2000, though overfishing means that many fisheries are now in danger of collapse.

Improved information technologies made it easier to accumulate, store, keep track of, and use the huge amounts of information that drove innovation and kept hugely complex modern societies running. Communications and transportation technologies transformed collective learning by creating, for the first time, a single, linked network of minds that spanned the globe and could manage and track down new information in colossal electronic stores of information. The noösphere, the sphere of mind, became a dominant driver of change within the biosphere. Cheap but powerful networked computers gave billions of people access to more information than they could have found in all the libraries of the premodern world. When combined with the mathematically sophisticated techniques of modern statistical analysis, computers allowed governments, banks, corporations, and individuals to keep track of huge flows of resources. They also allowed instant communication between individuals anywhere in the world through telegrams, phones, and the Internet. If the sharing of information is what makes us humans so powerful, computers multiplied that power many times over. As always, there were losses, too. Just as memory skills probably declined with the spread of writing, so calculating skills declined with the spread of computers and calculators.

By 2000, the fossil-fuels revolution embraced most of the world, including many older hub regions. The yawning gaps in national wealth and power of the late nineteenth century began to close. European powers, weakened by the world wars, grudgingly gave up their colonies, and older hub regions in Asia, the eastern Mediterranean, North Africa, and the Americas began to catch up in technology, wealth, and power.

Behind all these changes was the bonanza of cheap energy from fossil fuels. Coal production increased everywhere, but so did the production of oil and natural gas. New oil fields were developed in Arabia, Iran, the Soviet Union, and even along the continental shelves. In the Middle East alone, oil production increased from 28 billion barrels in 1948 to 367 billion barrels in 1972, just twenty-five years later. Natural gas came into its own during the Great Acceleration. Total energy consumption doubled in the nineteenth century and then rose by ten times in the twentieth century. Human consumption of energy rose much faster than human populations.

Transforming the World: Governance and Society

The very nature of society and government was transformed by the new energy flows and technologies of the Anthropocene. Once, all humans had been foragers, and *government* really meant family relationships. After farming appeared, more and more people lived in peasant villages and supported themselves by farming. In farming societies, *government* meant, above all, mobilizing energy and resources from peasants. Today, most humans no longer gather or farm to produce their food and other necessities. They have become wage earners. Like the potters of ancient Sumer, they live on wages earned by doing specialized work. And that transformed the nature of government, because now governments had to become involved in the day-to-day lives of all their citizens. This is because wage earners, unlike peasants, cannot survive without governments. Farming villages could exist quite happily beyond the borders of the great agrarian civilizations, but wage earners depend on the existence of laws, markets, employers, shops, and currencies. A specialist wage earner, like a nerve cell, cannot survive alone. This is why a world of wage earners is much more tightly integrated than a world of peasant farmers. Modern governments regulate markets and currencies, protect the businesses that provide employment, create mass educational systems that can spread literacy to most of the population, and provide the infrastructure for the movement of goods and workers. To do all this, they have to draw more and more of their subjects into the work of government and administration.

We can see the changeover to modern types of government in the nineteenth century, as industrialization took off, more and more peasants became wage workers, and governments began to mobilize more of their populations. Revolutionary France, transformed by revolution and under attack from most of Europe, was one of the first modern states to recruit soldiers systematically from the entire population. The government of the United States was also forged in a period of war during which it had to mobilize much of the population. To do that, governments needed detailed records on the number of citizens, on their health and fitness, on their education, skills, wealth, and loyalty. These were problems most traditional governments had been able to ignore. The governments of revolutionary France and the United States began to mobilize the loyalty of their subjects through democratization, which brought more of the population into the work of government, and through nationalism, which appealed to people's sense of a shared national community. They offered increasing numbers of their subjects (wealthy men, other men, and women, in that order) some role in government through elections. Through schools and the rapidly developing news media, governments tried to reach into the minds of their subjects and generate new forms of loyalty. Nationalism proved a powerful way of uniting people with different traditions, religions, and even languages. It mobilized traditional instincts of kinship by constructing in the minds of citizens a vast, imagined family of millions of people to whom they owed loyalty, service, and, in the extreme crises of war, perhaps even their lives.

The total wars of the early twentieth century turned

governments into economic managers, as they tried to mobilize all the people and resources of modern industrial economies. We can roughly track the increasing role of government in economic management. In the late nineteenth century, the French government accounted for about 15 percent of French GDP, a very rough measure of total national production. At the time, that seemed like a lot. Contemporary governments in Britain and the United States accounted for less than 10 percent of their GDP. The wars of the early twentieth century forced governments to intervene more actively in economic management, and by the middle of the twentieth century, their economic role had increased everywhere. In the early twenty-first century, the average share of national expenditure controlled or managed by governments in the countries of the OECD (Organisation for Economic Co-Operation and Development, founded in 1960) was 45 percent of GDP, with most richer countries falling in the range from 30 to 55 percent.[4] Some governments, such as the Communist regimes of the Soviet Union and China, attempted to micromanage the entire national economy. Modern governments also wielded coercive power on a much larger scale than traditional governments had, through armies and police equipped with modern weaponry. Such power would have been unimaginable to the author of the *Arthashastra,* the ancient Indian treatise on statecraft. Modern governments have a scale, reach, power, and heft that make even the most powerful governments of the agrarian era look like featherweights.

In an increasingly interconnected world, governance also assumed more global forms. By the late twentieth century, there were many political structures—not yet governments—that managed, advised, and administered on a global scale. They included the United Nations, the International Monetary Fund, and large numbers of corporations and nongovernmental organizations (NGOs) such as the Red Cross, whose activities range across many different countries. These institutions represent, in embryonic form, a new, global level of governance that would have been unimaginable just a few centuries ago.

New Ways of Living and Being

Technological and political transformations have been accompanied by equally radical changes in human lifestyles—in the *experience* of life.

Modern humans live in ways that would have baffled, confused, and possibly terrified our ancestors. All the many different activities of a peasant household—plowing, sowing, harvesting, feeding livestock, milking cattle, cutting firewood, gathering mushrooms or herbs, bearing and rearing children, cooking the foods and weaving the fibers you have grown—dominated the lives of most people for thousands of years. Today, most farmers are entrepreneurs or wage earners. They work on huge industrial farms that specialize in just a few crops, some of them genetically engineered. They cultivate and transport their crops using lashings of fertilizers and pesticides and energy-hungry harvesters, tractors, and trucks. Modern farmers grow crops not to eat but to sell. They manage businesses. They borrow money from banks and buy their seeds, fertilizers, and tractors from large corporations.

Most people no longer live in villages but in towns and cities. Away from the fields, streams, and woods of the peasant village, they live in environments almost entirely shaped by human activity. As different jobs and skills and forms of expertise proliferate, people spend more and more time learning. Information—expert knowledge—is what counts, rather than the generalized skills of peasants. Increasing numbers of people enjoy levels of nutrition and health that were rare even a century ago, thanks to the productivity of modern agriculture and modern advances in medicine and health care. Modern anesthesia has ended the agony of most traditional medical interventions. (No longer is an amputation or tooth extraction made easier to bear by nothing but a shot of liquor.) Perhaps most remarkable of all, in just a century, these changes have more than doubled the average life expectancy of human beings.

Despite the wars of the twentieth century, interpersonal relations have also become, for the most part, less violent. There is a clear logic to this change, as coercion has become a less effective way of controlling behavior in the last century or two (when did you last see a public flogging?), and economic rewards and punishments have slowly taken their place (you probably *have* asked for a pay raise). Though today most people take for granted that slavery and domestic violence are wrong, it is important to remember that, as late as the eighteenth century, the slave trade remained quite respectable in most of the world; torture and execution were standard punishments even for petty crimes and widely regarded as a form of public entertainment; and beatings or corporal punishment were regarded as a normal and perfectly acceptable way of maintaining order within families and schools. Personal violence is still all too common, but, relative to the number of people in the world, it is much rarer than it used to be and no longer regarded in most of the world as an acceptable way of controlling behavior.

In the world of peasants, most lived close to subsistence, periods of shortage were familiar and common, and affluence meant, for most people, a solid home, freedom from debt, and enough money to pay taxes and feed and clothe a family. Today's consumerist world is utterly different. It is fueled by economic systems that, in the more affluent parts of the world, produce so much material wealth that their very survival depends on massive, sustained consumption by a rapidly growing global middle class. The idea of progress, which most of us take for granted, is also new. For the majority of human history, people assumed that, barring catastrophes, children would live much as their parents had.

Attitudes toward families and children have changed profoundly. In recent centuries, improved nutrition and health care began to lower child mortality, so more children survived into adulthood. Yet traditional peasant attitudes ensured that families kept trying to produce as many children as possible. Such attitudes, along with increasing food production, high fertility, and declining mortality helped drive the extraordinarily rapid population growth of recent centuries. Eventually, though, traditional attitudes began to change as families moved into towns, as educating and rearing children became more expensive, and as more children survived to adulthood. Urban families began to have fewer children, and fertility rates began to fall. The fall in fertility rates after the earlier fall in mortality rates is what demographers call the *demographic transition:* the emergence of a new demographic regime of low fertility and low mortality. And that explains why, in the twentieth century, rates of population growth began to slow, first in more affluent countries, and then throughout the world. It also helps explain fundamental changes in gender roles. Reduced pressure on women to spend their entire adult lives bearing or rearing children blurred traditional divisions between male and female roles and allowed women to take up roles from which they had been excluded during most of the agrarian era.

For anyone alive today, these aspects of modern lifeways are familiar, though the contrast with the now-vanished world of the peasantry may be harder to appreciate. Even harder to grasp is the staggering increase in the complexity of modern societies, the way every detail of your life is enmeshed in networks involving millions of other people who supply food and employment, health care, education, electricity, the fuel for your car, the clothes you wear. Each of these chains of interconnection may include thousands or millions of other humans linked together in networks of fabulous complexity. In idle moments at airports, I like to try to calculate how many people are involved in the project of building and maintaining an Airbus 380 and getting it from Sydney to London. Weaken any of these links, and our worlds can break down terrifyingly fast, as is apparent today in those parts of the world where state structures have collapsed. Kautilya, the author of the *Arthashastra,* would have said that humans in these places live under "the law of the fish."

Transforming the Biosphere

The fossil-fuels revolution and the Great Acceleration did not just transform human societies; they are also transforming the biosphere. The activities of humans are changing the distribution and number of living organisms, altering the chemistry of the oceans and the atmosphere, rearranging landscapes and rivers, and unbalancing the ancient chemical cycles that circulate nitrogen, carbon, oxygen, and phosphorus through the biosphere.

It has taken researchers a long time to realize that the impact of human activities is now as great as that of the major biogeochemical processes that maintain the stability of the biosphere. Without really understanding what we are doing, we are fiddling with the biospheric thermostats that have kept Earth's surface within habitable temperatures for four billion years.

Carbon is central to the chemistry of life, and its distribution in the atmosphere, the sea, and the crust has helped determine temperatures at Earth's surface throughout the planet's history. Today, as we tap the energy in fossil fuels, we are pumping huge amounts of carbon dioxide back into the atmosphere. But not until the 1950s did scientists seriously consider the impact this might have on the carbon cycle. Charles Keeling began measuring levels of atmospheric carbon dioxide levels in Hawaii in 1958. Within a few years, he found that those levels were rising fast. Before the fossil-fuels revolution, human emissions of carbon dioxide were not large enough to affect the levels of atmospheric carbon dioxide. Today, though, human activities are releasing about ten thousand megatons of carbon dioxide into the atmosphere each year, and it is estimated that since the industrial revolution, the total emissions amount to about four hundred thousand megatons of carbon dioxide.[5] How significant these changes are became apparent when researchers found ways of measuring carbon dioxide levels over hundreds of thousands of years. One method was to study ice cores, which contain tiny bubbles, trapped year by year, that can tell us the composition of the atmosphere on geological time scales. These showed that, in the two centuries since the industrial revolution, levels of atmospheric carbon dioxide had risen to levels higher than any seen for almost a million years.

The changes Keeling noted were real; they were striking; and they were transforming the carbon cycle. Rising carbon dioxide levels will mean warmer climates, and warmer climates will mean more energetic hurricanes, storms, and wind currents and rising ocean levels that will flood low-lying cities. The effects will persist for many generations because, once released into the atmosphere, carbon dioxide stays there for a long time. But carbon dioxide is not the only important greenhouse gas whose atmospheric levels have increased as a result of human activities. Levels of methane have risen even faster in the past two centuries, driven largely by the spread of rice-growing in flooded fields and the increasing number of domestic livestock. Methane is an even more powerful greenhouse gas, though it breaks down faster.

In the late twentieth century, computers allowed climate scientists to build increasingly sophisticated models of the likely impact of such changes on the atmosphere. Their models suggest that, within a few decades, as greenhouse-gas emissions create a warmer world, melting glaciers and ice caps will raise sea levels, drowning many coastal cities, and increased heat energy and evaporation will ensure more erratic, unpredictable, and extreme weather patterns and make agriculture more difficult. Within a few decades, global climates will look very different from the relatively stable patterns of the Holocene. As one US climate scientist puts it: "The climate is an angry beast, and we are poking it with a stick."[6]

Nitrogen is as vital for life as carbon. In 1890, human impacts on the nitrogen cycle were insignificant. Each year, humans extracted about fifteen megatons of nitrogen from the atmosphere, mainly through farming, while wild plants extracted about one hundred megatons, or almost seven times as much. One hundred years later, humans and plants had swapped roles. By 1990, the area of farmed land had increased to such a degree that wild plants were extracting only about 89 megatons, while human extraction of nitrogen through farming and fertilizer production had risen to 118 megatons.

Our impact on other large mammals has also been profound. In 1900, wild land mammals accounted for the equivalent of about ten megatons of carbon biomass. Humans already accounted for about thirteen megatons, while domesticated mammals — our cows, horses, sheep, and goats — accounted for an astonishing thirty-five megatons. In the next century, these ratios would get even more warped. By 2000, the total biomass of wild land mammals had fallen to about 5 megatons, while that of humans had increased fast (not surprising, given what we know of population growth) to about 55 megatons and that of domesticated mammals to an astonishing 129 megatons. This is a powerful indicator of the extent to which expanding human activities have squeezed out other species of large animals by taking more and more of the biosphere's resources.

The point is a general one. *Most* species of animals and plants that are not of immediate value to humans are declining in numbers. They are declining so fast that some speculate that we may be witnessing the early stages of another mass-extinction event. Rates of extinction are now hundreds of times faster than in the past few million years and approaching rates not seen since the last mass-extinction event, sixty-five million years ago. We humans have even managed to drive our closest relatives to extinction, including, probably, our hominin relatives, such as the Neanderthals. Our closest living relatives, the chimpanzees, gorillas, and orangutans, are close to extinction in the wild.

The fossil-fuels revolution has magnified the scale of human impacts in many other areas. Mining, road building, and the spread of cities now move more earth than is moved by erosion and glaciation. Diesel pumps suck fresh water from aquifers ten times faster than natural flows can replenish them. We are producing minerals, rocks, and forms of matter that never existed before. They include plastics (made from oil, and now accumulating in landfills in cities and within the oceans), pure aluminum, stainless steel, and vast amounts of concrete, a human-made rock whose manufacture is now a major contributor to carbon emissions. Such a proliferation of new substances has not been seen on Earth since the appearance of an oxygen-dominated atmosphere, around 2.4 billion years ago.[7]

One of the most terrifying of these changes is the increasing productivity of human weaponry. Just a few centuries ago, our most lethal weapons were spears or, perhaps, rock-throwing catapults. From the late medieval age, the gunpowder revolution, pioneered in China, gave us muskets, rifles, cannons, and grenades. World War II spawned weapons that could degrade the entire biosphere in just a few hours, weapons with the destructive power of the asteroid that did in the dinosaurs.

Measuring Change in the Anthropocene

New flows of information and energy have woven humans, animals, and plants, as well as the chemicals of the earth, seas, and atmosphere, into a single system constructed primarily for the benefit of our own species. This system depends on huge flows of energy from fossil fuels. We can roughly measure the impact of these energy flows in the Anthropocene using figures in the statistical appendix.

The first thing that stands out is the sheer scale of change in recent centuries. In the past two hundred years, human populations (column B) rose from nine hundred million to more than six billion. That is the equivalent of adding twenty-six billion people in a thousand years, a rate of growth one thousand times faster than that of the agrarian era, in which, on average, about twenty-five million people were added each millennium. Such growth rates are unsustainable, and in recent decades, they have been slowing. Nevertheless, the figures illustrate the stunning impact on population growth of the fossil-fuels revolution.

Rapid population growth depended on huge increases in the energy available to our species (column C). In the eight thousand years between the end of the last ice age and two thousand years ago, human energy consumption increased by about seventy times. In just two hundred years, between 1800 and 2000, total energy consumption rose by about twenty-two times, from 20 million gigajoules (20 exajoules) to 52 million gigajoules (520 exajoules). That rise is the equivalent of an increase of 2,500 exajoules every thousand years, a rate of increase twenty thousand times as fast as in the agrarian era.

The energy bonanza from fossil fuels, like the energy bonanza from farming, paid for population growth, for the complexity taxes demanded by entropy, and, finally, for rising living standards, but on a much larger scale than in the agrarian era. And this time, the rise in living standards was not confined to a tenth of the human population but extended to a much larger emerging middle class.

Much of the energy bonanza from fossil fuels paid for increasing numbers of humans. It fed, clothed, and housed the five to six billion people added to the world's population in the past two centuries. But the fossil-fuels bonanza was so much greater than that from farming that a lot more was left over for other uses. We know this because column D shows that the energy available per person increased by almost eight times in the past one thousand years, while in the whole eight thousand years between the end of the ice age and two thousand years ago, it had less than doubled. In the past two hundred years, populations have grown at lightning speed, but energy flows have grown even faster.

A lot of the extra energy must have paid for the taxes demanded by entropy from increasingly complex societies. Much of that energy did no productive work or was dissipated as heat or in pollution or waste or the destruction of war. It was doing entropy's work of degrading complex structures. We have no good measures of the amounts involved, but they must be significant. Then there are the other complexity taxes, the energy and wealth that paid for the infrastructure of today's global societies. In the past two hundred years, the size of the largest cities rose from about one million (a level that had barely changed in two thousand years) to more than twenty million (column F). Given the infrastructure of electricity, sewers, roads, and public transport needed for a modern city and the challenges of policing and regulating the activities of twenty million people crowded into a small area, it is apparent that this represents a quantum leap in social and technological complexity. Complexity taxes pay for the construction and upkeep of buildings, buses, trains and ferries, sewers and roads; they pay for garbage collection, the electricity grid, law codes, policing, prisons and courts, and the links by ship, plane, train, and the Internet that bind cities throughout the world into a single network. Without these different systems, all driven by huge flows of energy, the complex structures of a modern city would break down fast. And cities, in turn, are linked by a complex infrastructure of highways, laws, and electronic communications to hundreds of thousands of smaller towns, villages, and isolated settlements. Though we have no way of measuring it precisely, we can be sure that complexity taxes account for a large share of the energy from fossil fuels.

But the bonanza from fossil fuels was so massive that a lot of energy was left over for one more task: that of improving human welfare. As in the agrarian era, a disproportionate amount of wealth still supports a tiny elite, so, as in the past, we can allocate a significant share of the energy bonanza to elite consumption. But so huge was the increase in energy and wealth that, for the first time in human history, consumption levels began to rise for a growing global middle class of billions of people, far more people than the entire population of the world at the end of the agrarian era. Thomas Piketty estimates that in modern European countries, 40 percent of the population controls between 45 percent and 25 percent of national wealth. The appearance of this middle class was a new phenomenon in human history. And more and more people are joining the new middle class as the numbers living in extreme poverty fall.

Paradoxically, increasing wealth also means increasing inequality, and even as the numbers living above subsistence are rising, the numbers living in extreme poverty remain higher than ever before in human history. Thomas Piketty estimates that in most modern countries, the wealthiest 10 percent of the population controls between 25 percent and 60 percent of national wealth, while the bottom 50 percent controls no more than 15 percent to 30 percent. This represents a decline in inequality in comparison with the era just before World War I.

But in the early twenty-first century, inequality seems to be on the rise again, and the huge number of people alive now means that, in absolute terms, there are far more people living in extreme poverty today than there were in the past. In 2005, more than three billion people (more people than the total population of the world in 1900) lived on less than $2.50 a day. Most people in this group have seen few benefits from the fossil-fuels revolution and suffer from the unhealthy, unsanitary, and precarious living conditions of the early industrial revolution that were described so vividly by Dickens and Engels.

Nevertheless, a growing proportion of the human population has benefited from increasing energy and wealth flows and is living well above subsistence. These flows have raised consumption levels and also levels of nutrition and health for billions of people. The measure that best captures this change is probably life expectancy (column E). For most of human history, life expectancies at birth were less than thirty years. This was not because people didn't live into their sixties and seventies but because so many children died young and so many adults died of traumas and infections that would not have killed them today. Life expectancies barely changed for one hundred thousand years. Then, in just the past one hundred years, average life spans have almost doubled throughout the world because humans have acquired the information and resources needed to care for the young and old much better, to feed more people, and to improve the treatment and care of the sick and injured.

The contrast between the energy bonanzas from fossil fuels and from farming is striking. The energy bonanza from fossil fuels was so vast that, in addition to expenditure on reproduction, elite wealth, waste, and the infrastructure for complexity, there was enough left over to raise the consumption levels and living standards of an increasing proportion of humanity. This was a revolutionary transformation. It occurred mostly in just the past one hundred years and primarily during the Great Acceleration of the second half of the twentieth century.

This is the face of the Good Anthropocene (*good* from a human perspective). The Good Anthropocene has generated better lives for billions of ordinary humans, for the first time in human history. (If you doubt the improvement, think again about having surgery without modern anesthesia.)

But there is also a Bad Anthropocene. The Bad Anthropocene consists of the many changes that threaten the achievements of the Good Anthropocene. First, the Bad Anthropocene has generated huge inequalities. Despite colossal increases in wealth, millions continue to live in dire poverty. And though it is tempting to think that the modern world has abolished slavery, the 2016 Global Slavery Index estimated that more than forty-five million humans today are living as slaves. The Bad Anthropocene is not just morally unacceptable. It is also dangerous because it guarantees conflict, and in a world with nuclear weapons, any major conflict could prove catastrophic for most of humanity.

The Bad Anthropocene also threatens to reduce biodiversity and undermine the stable climate system of the past ten thousand years. The flows of energy and resources that support increasing human consumption are now so huge that they are impoverishing other species and jeopardizing the ecological foundations on which modern society is built. In the past, coal miners took canaries into mines to detect carbon monoxide. Today, rising carbon dioxide levels, declining biodiversity, and melting glaciers are telling us that something dangerous is happening, and we should take notice.

The challenge we face as a species is pretty clear. Can we preserve the best of the Good Anthropocene and avoid the dangers of the Bad Anthropocene? Can we distribute the Anthropocene bonanza of energy and resources more equitably to avoid catastrophic conflicts? And can we, like the first living organisms, learn how to use gentler and smaller flows of resources to do so? Can we find global equivalents of the delicate proton pumps used to power all living cells today? Or will we keep depending on flows of energy and resources so huge that they will eventually shake apart the fantastically complex societies we have built in the past two hundred years?

PART IV

The Future

CHAPTER 12

Where Is It All Going?

It's tough to make predictions, especially about the future.

— YOGI BERRA (ATTRIBUTED)

Man has too long forgotten that the Earth was given to him for usufruct alone, not for consumption, still less for profligate waste.

— CHARLES PERKINS MARSH, *MAN AND NATURE*

Future Games

In the introduction we met the fantastic motley cavalcade of all things, with its stars and serpents, its quarks and cell phones, all marching to the distant thunder of supernovas under the unblinking but weary gaze of entropy. Where is the cavalcade going?

Oddly, few modern educational systems spend much time teaching systematically about the future. This neglect is surprising, because thinking about the future is something all brainy organisms do, and we humans do it better than any other species. Whether they belong to humans or chimps, brains create simplified models of the world as it is right now. They also create models of how the world could change. Brains, like stockbrokers and climatologists, are in the business of modeling futures. By doing so, they alert their owners to approaching possibilities and dangers.

Today, we humans can play future games with fantastic skill and on a fantastic scale. Our models are rich and powerful because human language and the sharing of information allow us to combine billions of individual models. That means we can refine, enrich, and improve our models as they are added to, tweaked, and corrected by feedback and new information from billions of other humans over many generations. Today's models of the world incorporate information from every part of planet Earth. We build them using the best of modern science and run them on networks of computers that can play through millions of different scenarios. "If all the glaciers in Greenland thaw, do sea levels rise enough to drown Miami and Dhaka?" That's a question we couldn't have asked seriously one hundred years ago. Today, rich and carefully tested answers to these kinds of questions can guide policy decisions that will affect billions of people, many of whom are young today, or not yet born. (And, yes, Miami and Dhaka would drown.)

Or we could ask much more grandiose questions about the remote future, such as "Will entropy win? Will it eventually break down *all* structures and forms?" As it happens, we have some pretty confident answers to such questions, because at cosmological scales, we are asking about relatively simple types of change. We're back with the complex physical systems of the early universe. Answers to cosmological questions about the future can't give us much practical guidance today because they are about events that are fantastically remote in time. But they can give shape to our modern origin story because they tease us with hints about where it is all going. They offer deep understanding, perhaps, and even a sense of closure, but not guidance.

Between the human and the cosmological time scales, there is another scale, of a few thousand years. What will the Earth look like in two thousand years? What will humans look like, for that matter? Or corncobs or cities or colonies on Mars?[1] Curiously, this in-between scale is the hardest to model. The interesting questions at this scale are about fantastically complex systems such as the biosphere, and in two thousand years, the tree of possibilities will have sprouted so many branches that even the most powerful computer models cannot pick the most likely. But it's not just the number of branches that stymies us. As quantum physics has shown, at the smallest scales, the universe is not deterministic. Unexpected things do happen, and, like the flapping of a butterfly's wings, they can cascade through causal chains with sufficient power to send the future off in many possible directions. So there is a lot of plain old-fashioned contingency. Neither our brains nor the best computer models can yet factor in a pandemic based on one tiny genetic mutation in a virus or the impact of a nearby supernova explosion, though we may be close to predicting a possible asteroid impact (knowledge the dinosaurs would have died for). At this intermediate scale, we enter the realm of science fiction. The stories we tell about the next few millennia are fascinating, haunting, and important. But we have no way of deciding which we should take seriously.

The Human Future: The Quest

For us humans, the next hundred years are really important. Things are happening so fast that, like the slow-motion time of a near accident, the details of what we do in the next few decades will have huge consequences for us and for the biosphere on scales of thousands of years. Like it or not, we are now managing an entire biosphere, and we can do it well or badly.

Myths of all kinds can tell us a lot about how to face an unpredictable future, because they are full of stories about near misses, catastrophic failures, and quests that succeeded. What's new today is a potential crash involving seven billion people, with millions of other organisms as bystanders and casualties. So, modern humans, like the heroes and heroines of all good myths, have a task. Our task is to avoid the crash and get to a good place for both humans and the biosphere, because we know there is no good place for humans in a ruined biosphere.

In the best myths, there are no guarantees. The crash really could happen. We could mishandle the intricate global machine we humans have built and lose the benefits of the Good Anthropocene. That is particularly likely if different drivers try to steer the machine in different directions or if we ignore the red warning lights appearing on its control panels. If the machine breaks down and productivity plummets, we won't be able to support seven billion people. We will face a grim period of social chaos, warfare, famine, and unchecked disease. This is the *Arthashastra*'s "law of the fish." If and when things finally settle, a much smaller number of survivors will be living once again within the energy limits of the agrarian era, in which only a tiny minority can enjoy more than a bare subsistence. If we do serious damage to climate systems, even agriculture may no longer work in much of the world. Farming depended, after all, on the stable climates of the Holocene.

Then, who knows? As in some science fiction, maybe remnant human populations will slowly rebuild something like our world, guided, perhaps, by memories and charred books and manuscripts or the broken-down vestiges of cities, factories, machines, and microchips. Or is it possible, as some have suggested, that there is a limit to the complexity we humans can manage? Have we reached a level of complexity that is simply beyond us? Is it, perhaps, the fate of all species capable of collective learning to hit a wall of complexity, at which point their societies collapse? Is that why we have not yet contacted any other species capable of collective learning? In the Greek myths, the gods punish Sisyphus, the king of Corinth, for being too clever and too ambitious. Advised, presumably, by entropy, they condemn him to push a boulder up a mountain and watch it roll down again, forever and ever.

These are bleak scenarios, but we cannot ignore them. The universe really is indifferent to our fate. It's a vast ocean of energy for which individual wavelets such as us are ephemeral, passing phenomena. "The hardness [of all great myths]," Joseph Campbell writes, "is balanced by an assurance that all that we see is but the reflex of a power that endures, untouched by the pain. Thus the tales are both pitiless and terrorless—suffused with the joy of a transcendent anonymity regarding itself in all of the self-centered, battling egos that are born and die in time."[2] Modern science captures the universe's terrifying indifference in the first and second laws of thermodynamics.

But we humans, like all living organisms, have goals, and we set out on long journeys to achieve those goals, despite the indifference of the universe. And stories from all cultures describe these dangerous journeys, journeys that don't always succeed but sometimes do. The journeyers endure periods when everything seems lost, periods of great suffering. There are sudden, unexpected interruptions to their quest. Helpers appear, too, gods or friends. And there are lucky breaks. So, in all mythological traditions, quests can and do succeed. Alertness, determination, and hope—these are the crucial virtues of anyone on a quest, because the journeyer who misses opportunities or who gives up too soon or who despairs must fail. Any traditional storyteller could have told us that these are the qualities we humans will need as we face an unpredictable future full of both dangers and opportunities.

Our discussion of the Good and the Bad Anthropocenes tells us what the goals of the human quest are right now. The first is to avoid a crash. If we can do that, there are two further goals: to ensure that the benefits of the Good Anthropocene are available to all humans, and to ensure that the biosphere continues to thrive, because if the biosphere fails, no quest can succeed. Our challenge is to achieve these goals, even if they often seem to point in different directions, sometimes toward indulgence, sometimes toward restraint.

Lest this sound too grandiloquent, here is how the human quest is described in the preamble to the United Nations document "Transforming Our World," published in 2015:

> All countries and all stakeholders, acting in collaborative partnership, will implement this plan. We are resolved to free the human race from the tyranny of poverty and want and to heal and secure our planet. We are determined to take the bold and transformative steps which are urgently needed to shift the world on to a sustainable and resilient path. As we embark on this collective journey, we pledge that no one will be left behind.

The document continues:

> People: We are determined to end poverty and hunger, in all their forms and dimensions, and to ensure that all human beings can fulfil their potential in dignity and equality and in a healthy environment.
>
> Planet: We are determined to protect the planet from degradation, including through sustainable consumption and production, sustainably managing its natural resources and taking urgent action on climate change, so that it can support the needs of the present and future generations.
>
> Prosperity: We are determined to ensure that all human beings can enjoy prosperous and fulfilling lives and that economic, social and technological progress occurs in harmony with nature.

There follow 17 sustainable-development goals and 169 specific targets that are to be achieved, if all goes well, over the next fifteen years.

It is easy to be skeptical. And some cynicism is appropriate. Nevertheless, for someone who grew up in the mid-twentieth century, when there was little understanding of the dangers of the Bad Anthropocene, it is remarkable to read such declarations from a body that represents most nations on Earth.

Soon after the sustainable-development goals were published, another landmark document appeared: the Paris Accord on Climate Change. This was adopted on December 12, 2015, at a UN conference attended by 195 nations. It came into force on November 4, 2016, when enough nations had formally ratified it. Its aims are as follows:

(a) Holding the increase in the global average temperature to well below 2°C above pre-industrial levels and to pursue efforts to limit the temperature increase to 1.5°C above pre-industrial levels, recognizing that this would significantly reduce the risks and impacts of climate change;

(b) Increasing the ability to adapt to the adverse impacts of climate change and foster climate resilience and low greenhouse gas emissions development, in a manner that does not threaten food production;

(c) Making finance flows consistent with a pathway towards low greenhouse gas emissions and climate-resilient development.

The tension between these two documents captures many of the difficulties of the quest for a better world, because it is really not clear that carbon dioxide emissions can be held to the declared targets without drastic cuts in use of fossil fuels. Are those cuts compatible with sustained growth? Perhaps, if renewable energy output increases rapidly enough. But the task would surely be eased if there were a greater commitment to redistribution and a willingness to accept slower rates of economic growth.

Our modern origin story suggests a helpful analogy, that of chemical activation energies. Activation energies provide the initial kick that gets vital chemical reactions going. But once they are under way, less energy is needed. Perhaps we can think of fossil fuels as the activation energy that was needed to kick-start today's world. Now that this glossy new world is in motion, can we keep it going with smaller and more delicate energy flows, like the tiny flows, electron by electron, or proton by proton, that are managed by enzymes and that energize living cells? Can we imitate respiration, big life's delicate, nondisruptive equivalent of fire?

The idea of fossil fuels as activation energy suggests something else about today's world. The turbulent dynamism of recent centuries is typical of all periods of creative destruction. It is the human equivalent of the gravitational energies that create stars. But once the violent energies of creation have done their work, we expect a new and more stable dynamism, as something new takes its seat in the universe. Like our sun, we can perhaps settle into a period of dynamic stability, having crossed a new threshold and built a new world society that preserves the best of the Good Anthropocene. Perhaps the idea of endless growth is completely wrong. Perhaps the disruptive dynamism of recent centuries is a temporary phenomenon. After all, living life within a framework of social and cultural stability has been the norm for most of human history and for most human societies. And that is why an understanding of what it means to live richly and dynamically in a less changeable world is preserved within the cultures of many modern indigenous communities whose people see themselves primarily as custodians of a world larger and older than themselves.

Though unfashionable at present, the idea of a future without continuous growth has popped up regularly in discussions by philosophically minded economists. Many eighteenth-century economists, including Adam Smith, feared a no-growth future, seeing it as the end of progress. But John Stuart Mill welcomed such a future as a refreshing contrast to the frenetic gold-rush world of the industrial revolution. In 1848, he wrote, "I confess I am not charmed with the ideal of life held out by those who think that the normal state of human beings is that of struggling to get on; that the trampling, crushing, elbowing, and treading on each other's heels, which form the existing type of social life, are the most desirable lot of human kind, or anything but the disagreeable symptoms of one of the phases of industrial progress."[3]

Instead, he argued, "the best state for human nature is that in which, while no one is poor, no one desires to be richer, nor has any reason to fear being thrust back, by the efforts of others to push themselves forward." Growth was still needed, he stated, in many poorer countries, but the richer countries were more in need of a better distribution of wealth. With basic necessities taken care of, the task for them was to live more fully rather than to keep acquiring more material wealth.

A stationary condition of capital and population implies no stationary state of human improvement. There would be as much scope as ever for all kinds of mental culture, and moral and social progress; as much room for improving the Art of Living, and much more likelihood of its being improved, when minds ceased to be engrossed by the art of getting on.

He warned that the stationary state should be chosen deliberately and on good terms before it was forced on a reluctant humanity on much poorer terms. "I sincerely hope, for the sake of posterity, that they will be content to be stationary, long before necessity compels them to it."

Many others have recognized that economic growth is not the same as a good life. In 1930, in an essay entitled "Economic Possibilities for Our Grandchildren," the British economist John Maynard Keynes argued that within a century, productivity would be high enough to guarantee the necessities of life to everyone. At that point, he hoped, people would stop working so hard and think more about how they lived. In March 1968, just before he was assassinated, Robert Kennedy described the limitations of an economy devoted to never-ending growth in gross national product:

> The Gross National Product counts air pollution and cigarette advertising, and ambulances to clear our highways of carnage....It counts the destruction of the redwood and the loss of our natural wonder in chaotic sprawl....Yet the GNP does not allow for the health of our children, the quality of their education, or the joy of their play. It does not include the beauty of our poetry or...the intelligence of our public debate or the integrity of our public officials....It measures everything, in short, except that which makes life worthwhile.

Our growing understanding of the biosphere tells us why we need to treat it more gently. How resilient is the biosphere, after all? We don't really know. There may be tipping points that will accelerate damaging changes by setting in motion dangerous positive-feedback cycles. For example, glaciers, such as those that cover most of Greenland, reflect sunlight. When they melt, Earth turns darker and begins to absorb heat instead of reflecting it. This increases the amount of heat retained in the atmosphere, and that melts more glaciers, which reduces Earth's reflectivity, which increases warming even further. Such mechanisms suggest why we need to think hard about biospheric limits.

The Stockholm Resilience Centre has worked for many years at identifying "planetary boundaries": limits humanity cannot cross without seriously endangering our future.[4] They have identified nine crucial boundaries of which two, climate change and declining biodiversity, are critical because if either one is breached seriously, it could drive the biosphere beyond stable limits.[5] Of course, modeling changes at global scales is still a rough-and-ready business. Sirens won't go off as we cross these boundaries. But, with due caution, researchers at the center conclude that we have already crossed the planetary boundary for biodiversity quite decisively, and we are approaching the boundaries for climate change. We have crossed critical boundaries in our impacts on flows of phosphorus and nitrogen, and we are also close to the boundaries in our use of land, particularly forests. We are beginning to see red warning lights on the control panels of the global machine we humans have built.

If, despite all the challenges, we humans are successful in our quest, what will a "mature Anthropocene" look like?[6] It will not be a perfect world, of course. But it is important that we try to imagine such a world as we try to build it. There are so many imponderables here that we cannot produce any sort of architect's sketch. Nevertheless, we can describe some of the main features of a world that preserves the best of the Good Anthropocene while avoiding the dangers of the Bad Anthropocene.

Population growth will slow, eventually, to zero, and perhaps start falling. Rates of population growth are already falling in most parts of the world, and in some regions, the absolute number of people is beginning to fall. There are many steps that could speed the process, including better health care for poor families and better education for women and girls in poorer countries. Many economists warn about the dangers of slowing population growth, but a biospheric perspective shows why continued population growth is unsustainable. In a mature Anthropocene, poverty will be largely eliminated by better welfare systems and checks on the accumulation of extreme wealth. As we have seen, in relative terms, extreme poverty is already in decline in much of the world. Eventually, as economic growth ceases to become the primary goal of governments, individuals will begin to value quality of life and leisure over increased income. With the support of governments, more and more people will drop out of extreme forms of the rat race. Catering to these people's needs will boost sectors of the economy that provide services rather than material goods. Education and science will become more important to governments as knowledge begins to replace material goods as a source of wealth and well-being. Ideas will change, too—ideas about what a good life looks like and about the goals of good government.

The world's economies will wean themselves off fossil fuels sometime later in this century. Production of renewable energy is already increasing fast, so this is not an unrealistic goal, though it will require more vigorous intervention by governments than is apparent at present. When combined with measures to capture atmospheric carbon dioxide, a reformed global energy regime may limit global warming to two degrees Celsius above preindustrial levels. Increasing efficiencies in the use of energy and materials will eventually reduce *total* energy consumption, and recycling of existing materials will reduce consumption of new minerals and resources almost to zero.

Innovations and changes in consumption patterns will be part of a larger transformation of agriculture that makes it less demanding of resources and more efficient. Scientific innovation will surely play a huge role here. Much will be invested in protecting biodiversity, wetlands, and fragile regions such as coral reefs or tundra environments.

As Mill wrote, a more stable world need not be a static world. Indeed, it will offer rich opportunities for new forms of art, expanded and enhanced social life, and new and less manipulative ways of engaging with the natural world. Here, modern societies will have a huge amount to learn from those who have preserved traditions from the past, from societies that lived for thousands of years in a more stable relationship with their surroundings. And is it unreasonable to hope that in such a world, even if average consumption of resources does not increase, the quality of life may improve for large numbers of people?

Many of the Goldilocks conditions for crossing this new threshold are already emerging. They include the staggering intellectual wealth of modern scientific scholarship, a much better understanding of how the biosphere works, and a growing awareness that we humans share a common fate on our one home, planet Earth. We will also need vivid images of a better future to motivate action today. Hope is, after all, a crucial virtue as we try to build a better world, as is alertness (lots of good science will help) and determination (politics will play a crucial role here).

As I write this in 2017, determination is the virtue that seems least present. It is remarkable that governments throughout the world now pay lip service to something like the quest I have described. But there is still not yet a strong global consensus about the quest. Many remain convinced that the flickering warning lights are caused by faulty switches and bad science. And few have the luxury of thinking on the grand scales needed to seriously imagine the near future. Most people, but particularly the very poor, have to concentrate on personal needs and goals. And most politicians and entrepreneurs have to focus on more immediate issues. Governments are national and they are competitive, which means that the wealth and power of each individual nation tends to loom larger in political calculations than the needs of the world as a whole. Most governments are also tied to short-term goals by the methods by which officials are chosen or elected. Few can set firm and realistic goals for twenty or thirty years in the future, yet these are the time frames that will decide the outcome of the quest for a better world. Finally, in a capitalist world, most enterprises are governed by the need to make profits, and at present profit-making all too often points in different directions from the quest for sustainability.

So what chance is there of an emerging global consensus on the importance of the quest? One of the most hopeful signs is the speed with which a scientific consensus has been reached, reflected in documents such as the UN sustainability goals and the Paris climate accords. Thirty years ago, such declarations would have been inconceivable. We may also be close to an economic tipping point at which the quest itself turns out to be profitable and compatible with an evolving global capitalism. If that happens, the colossal innovative and commercial energies of modern capitalism and the power of governments that depend on the wealth generated by capitalism may swing behind the quest and give it the sort of boost that capitalist governments gave to the industrial revolution. But today, in a more complex world, the behavior of governments will depend, in part, on the existence of voters who take the quest seriously. That will depend to some extent on how well and how persuasively people can describe the quest itself.

If we successfully manage the transition to a more sustainable world, a sort of threshold 9, it will become apparent that human history really constitutes a single threshold of increasing complexity culminating in the conscious management of an entire biosphere. We see human history in sections just because we are so close to it. The larger, combined threshold began with collective learning. Just as gravity concentrated clouds of matter in the early universe, collective learning generated denser and more complex human societies, accelerated change, and created new forms of dynamism by giving humans increasing control over the biosphere. Accelerating change could have continued indefinitely until it led to a catastrophic explosion — the human equivalent, perhaps, of a supernova. But if we successfully negotiate the transition to a sustainable world, it will look, in retrospect, as if we humans generated a new and more stable form of complexity, just as fusion generated the new and more stable structures of stars by pushing back against gravitational contraction. Then we will see that thresholds 6 to 9 have created a new type of biosphere on planet Earth, with new thermostats and new and more conscious forms of regulation embedded within the noösphere, the sphere of mind. What should we call that threshold? The Human Revolution?

Beyond Humans: Millennial and Cosmological Futures

Let's be optimistic and imagine a world in which the quest has succeeded. Threshold 9 has been successfully negotiated and most humans are flourishing within a stable global society based on a more sustainable relationship to the biosphere. That means human societies may be around for several thousand years, perhaps even for hundreds of thousands of years.

Speculating on what comes next takes us into the terrifying, unpredictable, but perhaps Utopian world of the middle future. At this scale, our models are really guesses. Their chances of being right are about as great as nineteenth-century pictures of aristocrats in checkered suits riding bicycles on the moon. The best we can do is run through a list of some possibilities based on trends we can already see.

Will we see the emergence of global governmental structures that partially supersede nation-states and finally eliminate the threat of nuclear war? Will fusion power provide a new energy bonanza? If so, will we use it with greater sensitivity to its disruptive impacts on the biosphere, as a tool that can lay the foundations for a good life for all humans? Or will we find ways of controlling even vaster flows of energy to create civilizations of unimaginable complexity? A Russian astronomer, Nikolai

Kardashev, has argued that if there are other civilizations capable of something like collective learning, many will have learned to capture all the usable energy of their home planets, while some may have learned to manage all the energy of their solar system, and some may even have learned to tap the energy of entire galaxies.

Will our descendants migrate beyond Earth? Will they start mining asteroids or setting up colonies on the moon or Mars? Or (if we look far enough ahead) on life-friendly planets around nearby star systems? Will we engineer new life-forms, new, energy-efficient food crops, or microbes that can treat diseases or check cancers? Will we engineer tiny machines, nano-surgeons, that can enter our bodies and fix broken organs, or build buildings without supervision as they follow electronic architects' designs? Will we build machines much cleverer than us? If so, can we be sure we will keep control of them?

Will we build new humans? Will micro- and macro-enhancements make us bionic, give us longer and healthier lives, and eventually turn us into something different, something trans-human? Will new technologies allow humans to exchange ideas, thoughts, emotions, and images instantaneously and continuously, creating something like a single, vast global mind? Will the noösphere partially detach itself from us humans and turn into a thin, unified layer of mind hovering over the biosphere? When, in all of this, will we decide that human history (as we understand it today) has ended because our species can no longer be described as *Homo sapiens*?

Will new science transform our understanding of ourselves and the universe, turning today's origin story inside out? Comparing today's origin stories with those of one hundred years ago suggests that this could happen very soon, and many times.

And of course, there are also the unknown unknowns that could switch future tracks in a second or two. Our science and technology may already be good enough to see asteroid impacts coming and perhaps do something about them. But there may be other unpredictable catastrophes, such as…encountering other life-forms. If we do meet them, will we peer at them through a microscope (or bionically enhanced eyes)? Or will they pick us up with huge tweezers, put us into vast petri dishes, and peer at *us* through microscopes?

It's a relief to turn to even larger scales where we can focus once more on relatively simple things such as planets, stars, galaxies, and the universe itself.

We can track the movements of tectonic plates, so we can guess roughly where the continents will be in one hundred million years. At present, it looks as if continental plates will regather in a new supercontinent that has already been dubbed Amasia because it will join Asia and the Americas. The ultimate fate of planet Earth will be decided by the evolution of the sun. Our sun will live for about nine billion years. But if it evolves like other, similar stars, in a few billion years it will start expanding and turning into a red giant. Earth will find itself inside the sun's outer layers. As Earth heats up, things will get tougher for big life, and there may be a long period in which the only survivors are tough archaebacteria, like those that survive in hot springs in Yellowstone Park. Eventually, even they will vanish as Earth is sterilized and then gobbled up and evaporated within the outer layers of an increasingly unstable and unpredictable red giant star. That's the end of planet Earth and of any still-living descendants unless they have traveled to the outer reaches of the solar system or to other star systems. As for the sun, after a long period as a red giant, it will eventually blow away its outer layers, turn into a white dwarf, migrate to the bottom of the Hertzsprung-Russell diagram, and then sit there, cooling, for hundreds of billions of years.

At about the time our sun goes rogue, our galaxy will collide with a neighboring galaxy, the Andromeda. This will be a sedate affair, like a crash between two clouds. But within each galaxy there will be a lot of turbulence as stars tug at one another in unpredictable ways. And the new, combined Milky Way/Andromeda galaxy will be a lot messier than the two beautiful spiral galaxies from which it was built.

What of the universe as a whole? Today, most cosmologists are pretty confident that there is a story to be told, because the future of the universe seems to depend on a small number of variables. The critical ones are the rate of expansion and the amount of matter/energy in the universe. It was once thought that the gravitational pull of matter in the universe would eventually rein in the expansion, put it into reverse, and shrink the universe down again into another primordial atom, which might in turn blow up and expand to create a new universe, and the sequence might be repeated in an infinite series of cosmological bounces. But since it was discovered in the late 1990s that the rate of expansion is increasing, it seems there must exist some kind of dark energy that is powerful enough to override the gravitational pull of all the mass and energy in the universe. That suggests that the universe will keep expanding forever and will do so faster and faster and faster.

As we talk about the remote future of the universe, we begin to realize that the story we've told so far was just the preface. The cavalcade of all things has a long and sometimes difficult journey ahead. We humans live right at the beginning of the universe's history, and its story is just getting going. Our universe is still young and energetic; it has plenty of living to do and plenty of complex new structures to build.

But in the very distant future, gazillions of years after we are all gone, the story gets darker, both literally and metaphorically. The universe will expand faster and faster, distant galaxies will vanish like ships over the horizon of space-time, and, eventually, anyone or anything left in our galaxy will feel seriously alone.[7] Stars will keep forming and burning until 10^{15} years in the future, when the universe is ten thousand times as old as it is today. By then it really will be showing its age, because the last stars will have stopped burning, and the lights will have gone out. Our galaxy will turn into a graveyard full of the cooling cinders of stars and planets.

But there will still be things moving in the graveyard. Black holes will slurp up the remains of the stars and planets. When they've finished doing that, they will turn on each other in cannibalistic civil wars until there are just a few huge bloated black holes left. These will sit there for unimaginable periods, perhaps for 10^{100} years, and will sweat energy until eventually they, too, will dwindle, fade away, and evaporate. It will turn out that everything that seemed permanent in our universe was actually ephemeral. Maybe even space and time will turn out to be mere forms, mere wavelets in a larger multiverse. Entropy will have finally destroyed all structure and order.

At least in one universe. But perhaps there are more to get working on.

Acknowledgments

I cannot possibly thank all the people who have helped me with this book by educating me, reading draft manuscripts, pointing me to important books and important authors, commenting on my lectures, and giving lectures that I have listened to. We humans swim in a sea of ideas, and a book like this is built by grabbing ideas as they float past, linking them to other ideas, bending them, perhaps warping them, perhaps finding new connections. I can trace some ideas to particular individuals and even particular conversations, but many got lodged in my brain and fermented, sometimes over several years, before turning up in another part of my brain in new forms and without labels to remind me of their sources. So I do not know whom to thank for many of the ideas in this book. All I can do is offer a general thank-you to my numerous colleagues and friends and to the rich process of collective learning that has stocked my mind with countless ideas from today's marvelous and prolific world. Big history is a collective project, an emergent property of synergies among many, many minds.

Some people I *can* thank directly. A small group of like-minded scholars have gathered around the idea of big history and its analogues and worked to advance education and research in big history. They include pioneers such as the astrophysicist Eric Chaisson, the sociologist Johan Goudsblom, and all those who helped form and nurture the International Big History Association (listed in alphabetical order): Walter Alvarez,

Mojgan Behmand (and her many colleagues at Dominican University), Craig and Pamela Benjamin, Cynthia Brown, Leonid Grinin, Lowell Gustafson, Andrey Korotayev, Lucy Laffitte, Jonathan Markley, John Mears (who began teaching big history at the same time as me), Alessandro Montanari, Esther Quaedackers, Barry Rodrigue, Fred Spier, Joe Voros, Sun Yue, and many others who have helped build the big history story. I worked particularly closely with Craig Benjamin and Cynthia Brown on the first college textbook in big history in a remarkably friendly and fruitful collaboration. Sadly, my friendship with Cynthia ended with her death on October 15, 2017; one of the pioneers of big history, she will be missed by everyone in the field. Many world historians have supported the idea of big history over the years, beginning with Felipe Fernández-Armesto, Bob Bain, Terry Burke, Ross Dunn, Pat Manning, Merry Wiesner-Hanks, and others. Two great world historians lent their immense prestige to the new field: William H. McNeill, who saw big history as the logical next phase beyond world history, and Jerry Bentley, who first invited me to publish on the relationship between big history and world history. The Teaching Company invited me to give a lecture series on big history, and Bill Gates, who listened to those lectures, gave a tremendous boost to the field by supporting the creation of a free online syllabus in big history for high schools and inviting me to give a TED Talk on big history in 2011. His support resulted in the Big History Project, very ably managed first by Michael Dix and colleagues from Intentional Futures and now by a team headed by Andy Cook and Bob Regan. Co-creators of the Big History Project include the hundreds of teachers and schools and students who have taken the courageous gamble of teaching and learning this ambitious new approach to the past. The World Economic Forum allowed me to speak about big history as a global project, and at the annual meetings in Davos, I have had the privilege of being introduced by two Nobel Prize winners: former vice president of the United States Al Gore and Australian astrophysicist Brian Schmidt. I also had the privilege of visiting Lake Mungo and meeting Mary Pappin, a Mutthi Mutthi elder whose family played a crucial role in the return to their homelands of the remains of Mungo Lady and Mungo Man.

I have spent most of my career at Macquarie University in Sydney, and Macquarie has supported the idea of big history since I first began teaching it, with colleagues from across the university, in 1989. My particular thanks to Bruce Dowton and his colleagues for supporting big history and the creation of the Macquarie University Big History Institute, very ably led by Andrew McKenna, Tracy Sullivan, and David Baker (who is, as far as I know, the first scholar to earn a PhD in big history). Over the years, my colleagues in the Modern History Department have given immense support to this new way of thinking about history, and many have taught big history alongside me. My thanks to all of them, and particularly to Marnie Hughes-Warrington, Peter Edwell, and Shawn Ross. My thanks also to my many students in big history, who kept me on the straight and narrow by always bringing me back to the simplest and deepest of questions. I spent a very enjoyable eight years at San Diego State University, whose historians provided both support and smart insights into how this new approach to history might play out in the diverse academic communities of the United States and whose graduate students proved to be remarkably disciplined and skilled tutors in big history.

Many experts in different fields have offered new insights or course corrections; they include Lawrence Krauss, Charles Lineweaver, Stuart Kauffman, Ann McGrath, Iain McCalman, Will Steffen, Jan Zalasiewicz, and many, many more. I have received immense support and rich feedback from my editors at Little, Brown, and at Penguin: Tracy Behar, Charlie Conrad, and Laura Stickney. I thank Tracy Roe for her scrupulous and eagle-eyed copyediting. And I owe a huge debt of gratitude to John

Brockman, who supported the idea of this book from the moment I first suggested it.

Several friends have kindly looked at and commented on drafts of this manuscript. They include Craig Benjamin, Cynthia Brown, Nick Doumanis, Connie Elwood, Lucy Laffitte, Ann McGrath, Bob Regan, Tracy Sullivan, and Ian Wilkinson.

For my family, big history has become something of a cottage industry. Chardi, Emily, and Joshua have all looked at drafts of this manuscript, and their comments and ideas over the years have often sent me in new directions. To Chardi, I owe the deep insight that big history is really a modern origin story. To them and my wider family (including my mother, who was my first teacher), I owe the deep gratitude of someone whose life has been blessed by the kindness and love of those closest to him. I dedicate this book to my family, to my grandchildren, Daniel Richard and Evie Rose Molly, and to all students everywhere as they embark on the momentous challenge of building a better world.

Appendix

Statistics on Human History

Statistics on Human History in the Holocene and Anthropocene Epochs*						
ERA	A: YEAR 0 = 2000 CE	B: POP. (Mill.)	C: TOTAL ENERGY USE Mill. GJ/Yr (=.001 Exajoules) (= B*D)	D: PER CAP ENERGY USE GJ/cap/Yr (1st 3 = max. est.)	E: LIFE EXPECTANCY (Years) 1st 3 = max. est.	F: LARGEST SETTLEMENT POP. (1,000s) 1st = max. est.
HOLOCENE	-10,000	5	15	3	20	1
	-8,000					3
	-6,000					5
	-5,000	20	60	3	20	45
	-2,000	200	1,000	5	25	1,000
	-1,000	300	3,000	10	30	1,000
ANTHRO- POCENE	-200	900	20,700	23	35	1,100
	-100	1,600	43,200	27	40	1,750
	0	6,100	457,500	75	67	27,000
	10	6,900	517,500	75	69	

* Columns A through E based on Vaclav Smil, *Harvesting the Biosphere*, loc. 4528, Kindle; column F based on Ian Morris, *Why the West Rules—for Now*, 148–49, plus 10,000 BP data interpolated.

Glossary

A list of technical terms or terms that are used in distinctive ways in this book.

absorption lines: Dark lines that appear when starlight is analyzed with a spectroscope; they indicate the presence of particular elements that have absorbed some of the energy of starlight and can be used to detect the motion of remote objects as the dark lines shift to the red or blue end of the spectrum.

accretion: The process by which matter in orbit around a star gathers together to form planets, moons, and asteroids.

activation energy: An initial shot of energy that initiates reactions that may generate much more energy, like a match starting a forest fire.

adaptive radiation: Periods of rapid biological evolution and diversification, often following mass-extinction episodes.

affluent foragers: Sedentary foragers such as the Natufians, usually found in regions of exceptional natural abundance.

agrarian civilizations: Communities of millions of people supported by agriculture with cities, states, bureaucracies, armies, social hierarchies, and writing.

agrarian era: The era of human history that was dominated by agricultural technologies; it started after the last ice age and ended two or three centuries ago.

agriculture: A suite of technologies that allowed humans to maximize the energy flows and resources available to them by manipulating the environment to increase production of plants and animals they found useful.

Anthropocene epoch: The most recent period of human history in which humans have become a dominant force for change in the biosphere; proposed as a new geological epoch, following on from the Holocene.

antimatter: Subatomic particles that are identical to other subatomic particles but have opposite charges, such as positrons (electrons with positive charges); when matter and antimatter meet, they obliterate each other and turn into pure energy.

arbitrage: Buying cheap on one market and selling dear on another market to make large profits.

archaea: Single-celled prokaryotic organisms; Archaea is one of the three major domains of life. *See also* bacteria *and* eukaryotes.

Archean eon: One of four major divisions in the history of planet Earth, from 4 billion years ago to 2.5 billion years ago.

astronomical standard candle: An astronomical object such as a Cepheid variable or a type 1a supernova whose distance can be determined, allowing it to be used to measure the distances to other objects.

atom: Smallest particle of ordinary matter, consisting of protons, neutrons, and electrons; atomic matter may account for only 5 percent of the mass of the universe. *See also* dark energy *and* dark matter.

ATP (adenosine triphosphate): Molecule used in all living cells to carry energy.

bacteria: Single-celled prokaryotic organisms in the domain of Eubacteria, one of the three major domains of life. *See also* archaea *and* eukaryotes.

big bang cosmology: Paradigm idea that arose in the 1960s to explain the emergence of our universe from a tiny, dense concentration of energy about 13.82 billion years ago.

biosphere: The sphere of planet Earth dominated and shaped by life and the by-products of living organisms.

black hole: A region so dense that nothing can escape its gravitational pull, not even light; often formed from the collapse of a supermassive star at the end of its life. There may be black holes at the center of all galaxies.

Cambrian explosion: The sudden proliferation of large organisms with hard body parts about 540 million years ago.

capitalism: A social system dominated by commercial activity and merchants in which governments favor commerce because much of their revenue comes from commerce.

Glossary

carbon: Element 6 on the periodic table; the fundamental element in living organisms because of the virtuosity with which it links up with itself and other elements.

catalyst: A molecule (usually a protein) that facilitates particular chemical reactions by lowering the required activation energy without the molecule itself being changed by the reaction.

Cepheid variable: A star whose brightness varies in a regular pattern. There are two main types, and because the rate of variation is related to their intrinsic brightness, their distance can be estimated so they can be used as astronomical standard candles to measure astronomical distances.

chemiosmosis: The movement of ions down their concentration gradient across a membrane. In cells, ATP synthase in the cell membrane harnesses this energy to charge up ATP molecules.

collective learning: The process, unique to humans, by which information is shared among individuals with such precision and in such volume that it accumulates from generation to generation; the key to our species' growing control of information and the biosphere.

complexity: Complex entities have more moving parts than simpler entities, and those parts are linked in precise ways that yield new emergent properties.

core: Central and densest region of Earth, dominated by iron and nickel; source of Earth's magnetic field.

cosmic microwave background radiation (CMBR): Radiation left over from the moment, about 380,000 years after the big bang, when the first atoms formed; still detectable today and one of the crucial pieces of evidence for big bang cosmology.

cosmology: Study of the universe and its evolution.

crust: The surface layer of Earth, made mostly of lighter rocks such as granites and basalts that have cooled sufficiently to solidify; this is where most organisms live.

dark energy: Energy whose nature and source is not yet understood but which may account for the accelerating expansion of the universe and for perhaps 70 percent of the mass of the universe.

dark matter: Matter whose gravitational effects are detectable but whose exact source and form is not yet understood; accounts for perhaps 25 percent of the mass of the universe.

demographic transition: In modern times, declining mortality drove population growth, but increasing urbanization eventually drove down fertility rates, so population growth is slowing today; the demographic transition has transformed the attitudes to families and gender roles that were dominant in most peasant societies.

differentiation: The process by which the early Earth heated up, melted, and was sorted into layers of decreasing density, among them the core, the mantle, and the crust.

DNA: Deoxyribonucleic acid, the molecule that carries the genetic information of most living organisms.

domestication: Genetic modification of a species as it coevolves with another species; fundamental to agriculture.

Doppler effect: Apparent change in frequency of emitted radiation as objects move toward or away from each other; used in police speed traps and to detect the motion of stars and galaxies toward or away from Earth.

Earth: The planet we live on, with its possibly unique cargo of living organisms.

electromagnetism: One of the four fundamental forms of energy. It is powerful at small scales, comes in positive and negative forms, and is the most important form of energy in chemistry and biology.

electron: Negatively charged subatomic particle; normally orbits atomic nuclei.

element: A basic form of atomic matter. Each element is distinguished by the number of protons in its nucleus; elements are classified within the periodic table according to their distinctive properties, and there are about ninety-two stable elements.

emergence: *See* emergent properties.

emergent properties: New properties that emerge as existing structures are linked together to form new structures with properties that are not present in their component parts. For example, stars have properties that are not present in the atomic matter from which they are constructed.

energy: The potential for things to happen or move or change. In our universe, energy comes in four main forms—gravity, electromagnetism, and the strong and weak nuclear forces—but it also exists in the form of dark energy.

entropy: The tendency of the universe to become less structured in accordance with the second law of thermodynamics.

enzyme: A biochemical molecule that acts as a catalyst, facilitating reactions in cells that would otherwise require much larger inputs of activation energy.

eukaryotes: Members of Eukarya, one of the three domains of life, eukaryotes are made up of cells with internal organelles. The first eukaryotes evolved through mergers between organisms from the other two (prokaryotic) domains of life, the Eubacteria and Archaea; all multicellular organisms consist of eukaryotic cells. *See also* archaea, bacteria, prokaryotes.

Fertile Crescent: The arc of well-watered lands around Mesopotamia in which agriculture first appeared.

fire-stick farming: Paleolithic technology based on regularly firing the land so as to increase its productivity.

first law of thermodynamics: *See* thermodynamics.

foraging: Characteristic technologies of the Paleolithic period, based on the gathering of resources from the environment and a limited amount of processing.

fossil fuels: Buried and fossilized organic material, primarily coal, oil, and natural gas, that contains ancient stores of energy from photosynthesis; the primary energy sources for the modern world.

free energy: Energy that does not flow randomly and so can do work (for example, the energy of water flowing through a turbine).

fusion: Occurs when protons collide so violently that they overcome the repulsion of their positive electric charges and are joined together by the strong nuclear force; fusion is accompanied by a huge release of energy as some matter is turned into energy. Source of the energy of H-bombs and the energy emitted by stars.

galaxy: A collection of millions or billions of stars held together by gravity; our home galaxy is the Milky Way.

gas: A state of matter in which individual molecules or atoms are not tightly bound together.

genome: The information stored in the DNA of every cell that regulates how it functions and allows it to make accurate copies of itself.

globalization: The increasing scale of exchange networks until, after 1500 CE, they began to reach around the entire world.

Goldilocks conditions: The rare special preconditions and environments that are "just right" to allow the emergence of new forms of complexity.

gravity: One of four fundamental forms of energy, though weak, gravity operates over large scales and tends to draw together everything with mass or energy. Einstein showed that gravity works by warping the geometry of space-time.

greenhouse gases: Gases such as carbon dioxide and methane that absorb and retain energy from sunlight; in sufficiently large quantities, greenhouse gases tend to raise temperatures at Earth's surface.

Hadean eon: One of four major divisions in the history of planet Earth; it began 4.6 billion years ago, when Earth first formed, and ended around 4 billion years ago.

half-life: The time it takes for half of a radioactive isotope to break down. Crucial concept for radiometric dating, as different half-lives allow different isotopes to be used to date events and objects at different time scales.

heat energy: The kinetic energy (or energy of motion) that drives the random jiggling of all particles of matter; only at a temperature of absolute zero does matter lose all heat energy. *See* temperature.

helium: Chemical element with atomic number 2 (two protons in its nucleus). Second most abundant element in the universe; chemically inert.

Hertzsprung-Russell diagram: Diagram charting the intrinsic brightness or luminosity of stars (the amount of energy they emit) against their color (or surface temperature); for astronomers, a powerful way of classifying different types of stars and the different ways in which stars evolve.

Holocene epoch: The geological epoch since the end of the last ice age, beginning about 11,700 years ago.

homeostasis: A state of equilibrium; living organisms maintain homeostasis by sensing changes in their environments and adjusting to those changes.

hominins: Bipedal apes that are ancestral to our own species and have evolved since our ancestors diverged from the evolutionary lineage leading to chimpanzees, about seven million years ago.

Homo sapiens: The species of great ape to which all readers of this book belong.

human: A member of the species *Homo sapiens*.

hydrogen: Chemical element with atomic number 1 (one proton in its nucleus); most abundant element in the universe.

ice ages: The era of ice ages interspersed with warmer interglacials that began about 2.6 million years ago, at the beginning of the Pleistocene epoch.

inflation: Cosmologically, a period of extremely rapid expansion of the universe early in the first second after the big bang.

information: The underlying rules that determine how change can occur. Some of these rules are universal, but living organisms need to be able to detect and react to *local information,* rules that work only in their immediate environment. *Information* can also refer to knowledge of how things work.

informavore: An entity that consumes information as carnivores consume meat; all living organisms are informavores.

isotope: Atoms of the same element with the same number of protons but different numbers of neutrons.

kelvin: Like the Celsius scale but begins at absolute zero (–273.15°C); the freezing point of water is 273.15 K and 0°C.

life: The emergent property of all living organisms. Hard to define precisely, as we only know of life on planet Earth, but its qualities include the ability to maintain homeostasis, metabolize, reproduce, and evolve.

light-year: Distance traveled by light in a vacuum during a single Earth year, approximately 9.5 trillion kilometers.

liquid: A fluid state of matter in which atoms or molecules are bound together but can flow past and around one another; liquid takes the shape of its container.

Luca: Last universal common ancestor; the inferred ancestor of all living organisms on Earth.

mantle: The semimolten layer of Earth beneath the crust and above the core, about three thousand kilometers thick.

map: In common usage, a schematic picture of a landscape or geographical region; often used here in a metaphorical sense to mean the pictures we create of space and time and of the entire universe and its history in order to identify our own place in the scheme of things.

matter: The physical "stuff" of the universe that occupies space. Einstein showed that matter consists of compressed energy and can be converted back to energy (for example, during proton fusion).

Glossary

megafauna: Large mammals; many were driven to extinction late in the Paleolithic soon after the arrival of humans in Australasia, Siberia, and the Americas.

metabolism: The ability of living organisms to tap and use energy flows from their environment.

metazoans: Multicellular organisms; "big life."

meteorite: A piece of space debris that lands on Earth; most meteorites have barely changed since the creation of the solar system so they provide information about the solar system's formation and evolution.

Milankovitch cycles: Variations in the orbit and tilt of Earth that affect the amount of energy it receives from the sun; these variations help explain the cycle of ice ages during the Pleistocene epoch.

molecule: Several atoms bound together by chemical bonds.

moon: The planetary body that orbits Earth, formed from a collision with another planetary body soon after Earth's formation.

multiverse: The speculative idea that there may be multiple universes, perhaps with slightly different fundamental laws and forms of energy.

Natufians: An archaeological term for "affluent foragers" who lived in the Fertile Crescent east of the Mediterranean between about 14,500 and 11,500 years ago.

natural selection: Charles Darwin's key idea that individual organisms survive and reproduce or fail to do so depending on how well they fit into their environments; this mechanism is the fundamental driver of evolution.

neutron: Subatomic particle usually found in atomic nuclei; similar mass as a proton but no electric charge.

nucleus: The dense core of an atom, mainly populated by neutrons and protons.

order (structure): Nonrandom or patterned arrangements of matter and energy.

origin story: An account of the evolution of all of space and time based on the best knowledge available to a particular community; origin stories are embedded within all major religious and educational traditions and provide a powerful way of understanding one's place in space and time.

oxygen: Chemical element, atomic number 8; fiercely reactive.

Paleolithic period: The era of human history from the initial appearance of our species, about two hundred thousand years ago, to the end of the last ice age and the beginning of farming, around eleven thousand years ago.

Pangaea: The supercontinent that existed from about three hundred million years ago to two hundred million years ago.

paradigm: An idea that is widely accepted by researchers in a particular field of study and that unifies information within that field; for example, big bang cosmology (astronomy), plate tectonics (geology), and natural selection (biology). Based on the work of the historian of science T. S. Kuhn.

parallax: The apparent movement of an object against its background as the observer moves; used by surveyors and astronomers to calculate distances to remote objects or nearby stars.

periodic table: Table of chemical elements, initially devised by Dmitry Mendeleyev, that groups elements with similar features.

Phanerozoic eon: One of four major divisions in the history of planet Earth, from about 540 million years ago to today; the era of large organisms or "big life."

phase change: A change of state, such as the change from the gaseous to the liquid or solid state.

photon: A massless particle of electromagnetic energy that moves at the speed of light in a vacuum and also has wavelike qualities.

photosynthesis: Capture of energy from sunlight by plants or plant-like organisms to power their metabolism.

planet: Astronomical body formed in orbit around a chemically enriched star.

plasma: A state of matter in which temperatures are so high that subatomic particles cannot bind together to form atoms.

plate tectonics: Paradigm idea that emerged in the 1960s to explain how convection currents within Earth's mantle, powered by heat in Earth's core, drive the movement of tectonic plates on Earth's surface.

Pleistocene epoch: Geological epoch from about 2.6 million years ago to about 11,700 years ago; dominated by ice ages.

prokaryotes: Single-celled organisms without nuclei, from the domains of Eubacteria and Archaea; the earliest life-forms on Earth were prokaryotes. *See* eukaryotes.

Proterozoic eon: One of four major divisions in the history of planet Earth, from around 2,500 million years ago to 540 million years ago.

proton: Subatomic particle with positive electrical charge found in atomic nuclei; the number of protons determines an element's atomic number.

quantum physics: The study of phenomena at the subatomic level, where it is impossible to identify the exact position or motion of particles, so physical laws have to be formulated as probabilities.

quark: Subatomic particle from which protons and neutrons are created by the strong nuclear force.

radioactivity: The tendency of many atomic nuclei to spontaneously break down, emitting subatomic particles.

radiometric dating: Dating techniques developed in the mid-twentieth century based on the regular breakdown of radioactive isotopes; this book's timeline could not have been constructed without radiometric dating techniques.

red giant: Dying star, such as Betelgeuse in Orion, that has expanded and has a cooler (redder) surface.

redshift: Shift of absorption lines toward the red end of the spectrum; an indication that an astronomical object is moving away from Earth. A key piece of evidence that the universe is expanding.

religion: Spiritual traditions, some highly institutionalized, all of which seem to have embedded within them some form of origin story.

respiration: The taking in of oxygen by animals; also, the use of oxygen in cells to release energy stored in sugars.

RNA: Ribonucleic acid, a close relative of DNA that is present in all cells and can both carry genetic information and do metabolic work.

science: Modern traditions of rigorous, evidence-based study of the world and the universe, developed since the seventeenth century's scientific revolution.

second law of thermodynamics: *See* thermodynamics.

sedentism: Nonnomadic lifeways, in which individuals and households mostly stay near their home base in permanent dwellings. Usually associated with agriculture but sometimes with affluent foragers.

solar wind: A flow of charged subatomic particles from the sun.

solid: A state of matter in which individual atoms and molecules are so tightly bound together that they cannot easily alter their position.

space-time: Einstein argued that space and time are best understood as part of a single universal framework, which he called space-time.

spectroscope: An instrument that breaks light into distinct frequencies; used to determine the chemical composition of astronomical objects.

star: An astronomical body formed when fusion reactions begin at the center of a collapsing body of matter; stars are gathered by gravity into galaxies.

strong nuclear force: One of four fundamental forms of energy. Operates at subatomic scales, binding quarks into protons and neutrons and holding atomic nuclei together.

subatomic particles: Components of atoms, such as protons, neutrons, and electrons.

sun: Our local star, source of most of the energy that powers the biosphere.

supernova: Massive explosion at the end of the life of a large star; many new chemical elements are generated within supernovas.

symbiosis: A relation of dependence between two species that is so close that they begin to affect how each species evolves; the human relationship to domesticated plants and animals is a form of symbiosis.

temperature: In scientific use, a measure of the average kinetic energy (or energy of motion) of the atoms from which something is composed.

thermodynamics: The study of how energy flows and changes form. The first law of thermodynamics asserts that the total amount of energy in the universe is fixed or "conserved"; the second law states that energy tends toward increasingly random or chaotic forms, so the long-term tendency of the universe is toward randomness or increasing entropy. *See* entropy.

thresholds of increasing complexity: Moments of transition when something new and more complex appears, with new emergent properties; the story told in this book is constructed around eight major thresholds of increasing complexity.

trophic level: Level in the food chain through which photosynthetic energy is transferred from plants to herbivores to carnivores and on to elites in human societies; significant amounts of energy are lost at each level so populations at higher levels are always smaller.

type 1a supernova: A type of supernova whose intrinsic brightness is known, so it can be used as an astronomical standard candle.

universe: The totality of all things of which we have evidence-based knowledge; formed in the big bang.

weak nuclear force: One of four fundamental forms of energy; acts at subatomic scales and responsible for many forms of nuclear decay.

white dwarf: Dense, dead star that has blasted away its outer layers and will cool down over many billions of years.

work: In thermodynamic theory, the ability to generate nonrandom change.

world zones: Large regions of the inhabited world (Afro-Eurasia, Australasia, the Americas, and the Pacific) that were almost entirely disconnected from one another before 1500 CE, so history evolved in distinct ways in each world region.

Further Reading

Notes indicate some of the books I have found most useful for particular topics. However, most works cited in the notes are recent accounts, and they do not include many classic texts that are now dated, such as H. G. Wells's *Outline of History* and Carl Sagan's wonderful *Cosmos*. The list below focuses mainly on books that train a wide-angle lens on the past, so it can be thought of as an introductory bibliography of works on big history and the modern origin story and books that take up some of the major themes in big history.

Books and Articles

Alvarez, Walter. *A Most Improbable Journey: A Big History of Our Planet and Ourselves*. New York: W. W. Norton, 2016. A personal exploration of the big history story by the geologist who showed that an asteroid did in the dinosaurs.

Brown, Cynthia Stokes. *Big History: From the Big Bang to the Present*. 2nd ed. New York: New Press, 2012. A version of the big-history story.

Bryson, Bill. *A Short History of Nearly Everything*. New York: Doubleday, 2003. A wonderful and highly readable account of the evolution of our modern scientific understanding of the universe.

Chaisson, Eric. *Cosmic Evolution: The Rise of Complexity in Nature*. Cambridge, MA: Harvard University Press, 2001. This book explores the link between energy-density flows and increasing complexity.

Christian, David. *Maps of Time: An Introduction to Big History*. 2nd ed. Berkeley: University of California Press, 2011. First published in

2004. One of the first modern attempts to tell the big-history story.

———. *This Fleeting World: A Short History of Humanity.* Great Barrington, MA: Berkshire Publishing, 2008. A short history of humanity.

———. "What Is Big History?" *Journal of Big History* 1, no. 1 (2017): 4–19, https://journalofbighistory.org/index.php/jbh.

Christian, David, Cynthia Stokes Brown, and Craig Benjamin. *Big History: Between Nothing and Everything.* New York: McGraw-Hill, 2014. A university textbook on big history.

Macquarie University Big History Institute. *Big History.* London: DK Books, 2016. A beautifully illustrated account of the big-history story.

Rodrigue, Barry, Leonid Grinin, and Andrey Korotayev, eds. *From Big Bang to Galactic Civilizations: A Big History Anthology, Vol. 1: Our Place in the Universe.* Delhi: Primus Books, 2015. An anthology of essays.

Spier, Fred. *Big History and the Future of Humanity.* 2nd ed. Malden, MA: Wiley-Blackwell, 2015. An ambitious attempt to tease out some of the main theoretical ideas behind big history.

Other Sources on Big History

Bill Gates has funded the creation of the Big History Project, a free, online big-history syllabus for high schools. Big history now has its own scholarly organization (the International Big History Association), and Macquarie University has established a Big History Institute to advance teaching and research in big history.

A TED Talk on big history that I gave in 2011 was designed to offer a short introduction to the idea of big history; it is available at https://www.ted.com/talks/david_christian_big_history.

Notes

I have tried to keep endnotes to a minimum except on topics where there is significant controversy.

Preface

1. William H. McNeill, "Mythistory, or Truth, Myth, History, and Historians," *American Historical Review* 91, no. 1 (Feb. 1986): 7.
2. H. G. Wells, *Outline of History: Being a Plain History of Life and Mankind*, 3rd ed. (New York: Macmillan, 1921), vi.
3. The great biologist E. O. Wilson has written eloquently about the vital importance of linking modern scholarly disciplines more closely; see E. O. Wilson, *Consilience: The Unity of Knowledge* (London: Abacus, 1998).
4. I first used that term in "The Case for 'Big History,'" *Journal of World History* 2, no. 2 (Fall 1991): 223–38.

Introduction

1. On the history of these finds and the very different perceptions of them by archaeologists and those who live today near Lake Mungo, see the wonderful short documentary by Andrew Pike and Ann McGrath, *Message from Mungo* (Ronin Films, 2014).
2. Superb on the archaeology of inland Australia is Mike Smith, *The Archaeology of Australia's Deserts* (Cambridge: Cambridge University Press, 2013).
3. *The Power of Myth,* episode 2, Bill Moyers and Joseph Campbell, 1988, http://billmoyers.com/content/ep-2-joseph-campbell-and -the-power-of-myth-the-message-of-the-myth/.

4. Alvarez, *A Most Improbable Journey*, 33.
5. In Fritjof Capra and Pier Luigi Luisi, *The Systems View of Life: A Unifying Vision* (Cambridge: Cambridge University Press, 2014), 280.
6. The Goldilocks principle has been explored thoroughly in Spier, *Big History*, 63–68 and following.

Chapter 1. In the Beginning: Threshold 1

1. Richard S. Westfall, *The Life of Isaac Newton* (Cambridge: Cambridge University Press, 1993), 259. Newton later changed his mind about the idea of the universe as God's "sensorium" but preserved the notion that God was "omnipresent in the literal sense."
2. Bertrand Russell, "Why I Am Not a Christian," lecture given at Battersea Town Hall, London, March 1927.
3. Cited in Christian, *Maps of Time*, 17.
4. Deborah Bird Rose, *Nourishing Terrains: Australian Aboriginal Views of Landscape and Wilderness* (Canberra: Australian Heritage Commission, 1996), 23.
5. Joseph Campbell, *The Hero with a Thousand Faces*, 2nd ed. (Princeton, NJ: Princeton University Press, 1968), 261.
6. Stephen Hawking, *A Brief History of Time: From the Big Bang to Black Holes* (London: Bantam, 1988), 151.
7. My thanks to Elise Bohan for this quote from Terry Pratchett, *Lords and Ladies* (London: Victor Gollancz, 1992).
8. On paradigms, the classic text is Thomas Kuhn, *The Structure of Scientific Revolutions*, 2nd ed. (Chicago: University of Chicago Press, 1970).
9. Peter Atkins, *Chemistry: A Very Short Introduction* (Oxford: Oxford University Press, 2015), loc. 722, Kindle.
10. Lawrence Krauss, *A Universe from Nothing: Why There Is Something Rather than Nothing* (New York: Simon and Schuster, 2012).
11. Erwin Schrödinger, *What Is Life?* And *Mind and Matter* (Cambridge: Cambridge University Press, 1967), 73.
12. Campbell, *The Hero with a Thousand Faces*, 25–26.
13. Peter M. Hoffmann, *Life's Ratchet: How Molecular Machines Extract Order from Chaos* (New York: Basic Books, 2012), loc. 179, Kindle.
14. For more on that idea, see Krauss, *A Universe from Nothing*.

Chapter 2. Stars and Galaxies: Thresholds 2 and 3

1. "From a molecular viewpoint, the raising of a weight corresponds to all its atoms moving in the same direction.... Work is the transfer of energy that makes use of the uniform motion of atoms in the surroundings." Peter Atkins, *Four Laws That Drive the Universe* (Oxford: Oxford University Press, 2007), 32.
2. See Chaisson, *Cosmic Evolution,* and Spier, *Big History.*
3. Andrew King, *Stars: A Very Short Introduction* (Oxford: Oxford University Press, 2012), 49.
4. Ibid., 59.
5. Ibid., 66.

Chapter 3. Molecules and Moons: Threshold 4

1. Peter Atkins, *Chemistry: A Very Short Introduction* (Oxford: Oxford University Press, 2015), loc. 788, Kindle.
2. Robert M. Hazen, "Evolution of Minerals," *Scientific American* (March 2010): 61.
3. John Chambers and Jacqueline Mitton, *From Dust to Life: The Origin and Evolution of Our Solar System* (Princeton, NJ: Princeton University Press, 2014), 7.
4. Doug Macdougall, *Why Geology Matters: Decoding the Past, Anticipating the Future* (Berkeley: University of California Press, 2011), 4.
5. ———, *Nature's Clocks: How Scientists Measure the Age of Almost Everything* (Berkeley: University of California Press, 2008), 58–60.
6. Tim Lenton, *Earth Systems Science: A Very Short Introduction* (Oxford: Oxford University Press, 2016), loc. 1297, Kindle.

Chapter 4. Life: Threshold 5

1. Both the metaphors and the calculations here come from Peter Hoffmann, *Life's Ratchet: How Molecular Machines Extract Order from Chaos* (New York: Basic Books, 2012), loc. 238, Kindle.
2. John Holland, *Complexity: A Very Short Introduction* (Oxford: Oxford University Press, 2014), 8. Complex adaptive systems contain "elements that are not fixed. The elements, usually called

agents, learn or adapt in response to interactions with other agents."

3. Seth Lloyd, *Programming the Universe* (New York: Knopf, 2006), 44.

4. Gregory Bateson, cited in Luciano Floridi, *Information: A Very Short Introduction* (Oxford: Oxford University Press, 2010), loc. 295, Kindle.

5. Daniel C. Dennett, *Kinds of Minds: Towards an Understanding of Consciousness* (London: Weidenfeld and Nicolson, 1996), 82.

6. David S. Goodsell, *The Machinery of Life,* 2nd ed. (New York: Springer Verlag, 2009), loc. 700, Kindle.

7. "Any process that generates structure increases the latent information inherent in that structure, which corresponds to a decrease in entropy (reduced number of microstates)." From Anne-Marie Grisogono, "(How) Did Information Emerge?," in *From Matter to Life: Information and Causality,* ed. Sara Imari Walker, Paul C. W. Davies, and George F. R. Ellis (Cambridge: Cambridge University Press, 2017), chapter 4, Kindle.

8. Hoffmann, *Life's Ratchet,* loc. 3058, Kindle.

9. Charles Darwin, *The Origin of Species* (New York: Penguin, 1985), 130–31.

10. The power of Darwin's idea and its capacity to shock are described superbly in Daniel Dennett, *Darwin's Dangerous Idea: Evolution and the Meaning of Life* (London: Allen Lane, 1995).

11. There is a good discussion of the Goldilocks conditions for rich chemistry in Jeffrey Bennett and Seth Shostak, *Life in the Universe,* 3rd ed. (Boston: Addison-Wesley, 2011), chapter 7.

12. Daniel C. Dennett, *From Bacteria to Bach: The Evolution of Minds* (New York: Penguin, 2017), 48.

13. *Science* 356, no. 6334 (April 14, 2017): 132.

14. Robert M. Hazen, "Evolution of Minerals," *Scientific American* (March 2010): 58.

15. Peter Ward and Joe Kirschvink, *A New History of Life: The Radical New Discoveries About the Origins and Evolution of Life on Earth* (London: Bloomsbury Press, 2016), 65–66.

16. Allen P. Nutman et al., "Rapid Emergence of Life Shown by Discovery of 3,700-Million-Year-Old Microbial Structures," *Nature* 537 (September 22, 2016): 535–38, doi:10.1038/nature19355.

17. Nadia Drake, "This May Be the Oldest Known Sign of Life on Earth," *National Geographic,* March 1, 2017, http://news.national

geographic.com/2017/03/oldest-life-earth-iron-fossils-canada
-vents-science/?WT.mc_id=20170606_Eng__bhptw&WT.tsrc=
BHPTwitter&linkId=38417333.

18. Madeline C. Weiss et al., "The Physiology and Habitat of the Last
Universal Common Ancestor," *Nature Microbiology* 1, article no.
16116 (2016), doi:10.1038/nmicrobiol.2016.116.

19. Nick Lane, *Life Ascending: The Ten Great Inventions of Evolution*
(New York: W. W. Norton, 2009), loc. 421, Kindle.

20. Terrence Deacon describes this as an autocell; see Grisogono,
"(How) Did Information Emerge?"

Chapter 5. Little Life and the Biosphere

1. On the idea of the biosphere, see Vaclav Smil, *The Earth's Biosphere:
Evolution, Dynamics, and Change* (Cambridge, MA: MIT Press,
2002), and Vladimir Vernadsky's pioneering work *The Biosphere*
(Göttingen, Germany: Copernicus, 1998), with a foreword by
Lynn Margulis. For a short summary of the history of the bio-
sphere, see Mark Williams et al., "The Anthropocene Biosphere,"
Anthropocene Review (2015): 1–24, doi: 10.1177/2053019615591020.

2. Christian, Brown, and Benjamin, *Big History,* 46.

3. Andrea Wulf, *The Invention of Nature: The Adventures of Alexander
von Humboldt, the Lost Hero of Science* (London: John Murray, 2015),
loc. 2368, Kindle.

4. Jeffrey Bennett and Seth Shostak, *Life in the Universe,* 3rd ed. (Boston: Addison-Wesley, 2011), 130.

5. Robert M. Hazen, "Evolution of Minerals," *Scientific American*
(March 2010): 63.

6. Bennett and Shostak, *Life in the Universe,* 134.

7. David Grinspoon, *Earth in Human Hands: Shaping Our Planet's
Future* (New York: Grand Central Publishing, 2016), 204.

8. See the discussion of these mechanisms in ibid., 44 and following.

9. Peter Ward and Joe Kirschvink, *A New History of Life: The Radical
New Discoveries About the Origins and Evolution of Life on Earth* (London: Bloomsbury Press, 2016), 64.

10. Dennis Bray, *Wetware: A Computer in Every Living Cell* (New Haven,
CT: Yale University Press, 2009), loc. 1084, Kindle.

11. Description from Gerhard Roth, *The Long Evolution of Brains and
Minds* (New York: Springer, 2013), 70.

12. See Andrew Knoll, *Life on a Young Planet: The First Three Billion Years of Evolution on Earth* (Princeton, NJ: Princeton University Press, 2003), 20; the book is superb on the staggering diversity of prokaryotic metabolic systems. On the energy flows tapped by the earliest organisms, see Olivia P. Judson, "The Energy Expansions of Evolution," *Nature: Ecology and Evolution* 28 (April 2017): 1–9.

13. Tim Lenton, *Earth Systems Science: A Very Short Introduction* (Oxford: Oxford University Press, 2016), 18.

14. Ibid., loc. 1344, Kindle.

15. Robert M. Hazen, "Evolution of Minerals," *Scientific American* (March 2010): 63.

16. Lenton, *Earth Systems Science,* loc. 1418, Kindle.

17. Donald E. Canfield, *Oxygen: A Four Billion Year History* (Princeton, NJ: Princeton University Press, 2014), loc. 893, Kindle.

18. Lenton, *Earth Systems Science,* loc. 1438, Kindle.

19. Roth, *The Long Evolution of Brains and Minds,* 73–75.

Chapter 6. Big Life and the Biosphere

1. Michael J. Benton, *The History of Life: A Very Short Introduction* (Oxford: Oxford University Press, 2008), loc. 766, Kindle, and see Dennis Bray, *Wetware: A Computer in Every Living Cell* (New Haven, CT: Yale University Press, 2009), loc. 2008 and following, Kindle.

2. Siddhartha Mukherjee, *The Gene: An Intimate History* (New York: Scribner, 2016), loc. 5797, Kindle.

3. Sean B. Carroll, *Endless Forms Most Beautiful: The New Science of Evo Devo and the Making of the Animal Kingdom* (London: Weidenfeld and Nicolson, 2011), 71 and following.

4. Much of the discussion that follows is based on Peter Ward and Joe Kirschvink, *A New History of Life: The Radical New Discoveries About the Origins and Evolution of Life on Earth* (London: Bloomsbury Press, 2016), chapter 7.

5. Doug Macdougall, *Why Geology Matters: Decoding the Past, Anticipating the Future* (Berkeley: University of California Press, 2011), 132.

6. Ward and Kirschvink, *A New History of Life,* 119.

7. Ibid., 124.

8. Niles Eldredge and S. J. Gould, "Punctuated Equilibria: An Alternative to Phyletic Gradualism," in *Models in Paleobiology,* ed. T. J. M. Schopf (San Francisco: Freeman Cooper, 1972), 82–115.

9. A wonderful, if controversial, book on the Burgess Shale fossils is Stephen Jay Gould, *Wonderful Life: The Burgess Shale and the Nature of History* (London: Hutchinson, 1989).

10. The term used by Ward and Kirschvink, *A New History of Life,* 222.

11. Tim Lenton, *Earth Systems Science: A Very Short Introduction* (Oxford: Oxford University Press, 2016), 44.

12. Ibid., 48: "The most pronounced change in atmospheric CO_2 over Phanerozoic time was due to plants colonizing the land. This started around 470 million years ago and escalated with the first forests 370 million years ago. The resulting acceleration of silicate weathering is estimated to have lowered the concentration of atmospheric CO_2 by an order of magnitude and cooled the planet into a series of ice ages in the Carboniferous and Permian Periods."

13. Ibid., 72.

14. Ibid., 24, on the relationship between carbon burial and atmospheric oxygen levels. Robert M. Hazen, "Evolution of Minerals," *Scientific American* (March 2010): 58, argues that by four hundred million years ago, Earth had its full complement of over four thousand types of minerals.

15. Gerhard Roth, *The Long Evolution of Brains and Minds* (New York: Springer, 2013), 229.

16. Daniel Cossins, "Why Do We Seek Knowledge?," *New Scientist* (April 1, 2017): 33.

17. The neuroscientist Antonio Damasio, in *Self Comes to Mind: Constructing the Conscious Mind* (Calgary, Alberta: Cornerstone Digital, 2011), argues that our sense of awareness is embedded within these constantly shifting maps of reality that begin with sensual, visual, and feeling maps of our own bodies.

18. Dylan Evans, *Emotion: A Very Short Introduction* (Oxford: Oxford University Press, 2001), loc. 334, Kindle.

19. Roth, *The Long Evolution of Brains and Minds,* 15–16.

20. Ibid., 162–63.

21. In this discussion, I'll be following closely the description of this event by Walter Alvarez, the geologist who demonstrated that an asteroid impact wiped out the dinosaurs; see his wonderful short book *T. Rex and the Crater of Doom* (New York: Vintage, 1998).

22. *Science News,* https://www.sciencenews.org/article/devastation-detec tives-try-solve-dinosaur-disappearance.

23. Stephen Brusatte and Zhe-Xi Luo, "Ascent of the Mammals," *Scientific American* (June 2016): 20–27.

24. Ward and Kirschvink, *A New History of Life,* 315.

25. Ibid., 316.

Chapter 7. Humans: Threshold 6

1. This is argued eloquently in David Grinspoon, *Earth in Human Hands: Shaping Our Planet's Future* (New York: Grand Central Publishing, 2016).

2. Robin Dunbar, *The Human Story: A New History of Mankind's Evolution* (London: Faber and Faber, 2004), 71.

3. Gerhard Roth, *The Long Evolution of Brains and Minds* (New York: Springer, 2013), 226.

4. It's an old joke. I came across it in Daniel Dennett, *Consciousness Explained* (London: Penguin, 1991), 177; Dennett attributes the comparison with tenure to the Colombian-American neuroscientist Rodolfo Llinás.

5. On this last idea, see Michael S. A. Graziano, *Consciousness and the Social Brain* (Oxford: Oxford University Press, 2013).

6. The complexities of ape and monkey politics are explored in works by Frans de Waal and Jane Goodall and more recently in a study of baboon communities by Dorothy L. Cheney and Robert M. Seyfarth, *Baboon Metaphysics: The Evolution of a Social Mind* (Chicago: University of Chicago Press, 2007).

7. See Christopher Seddon, *Humans: From the Beginning* (New York: Glanville Books, 2014), 42–45.

8. On EQ, see ibid., 225 and following, and Roth, *The Long Evolution of Brains and Minds,* 232.

9. Roth, *The Long Evolution of Brains and Minds,* 228.

10. See John Gowlett, Clive Gamble, and Robin Dunbar, "Human Evolution and the Archaeology of the Social Brain," *Current Anthropology* 53, no. 6 (December 2012): 695–96, on the correlation of brain size and group size.

11. *New Scientist* (April 29, 2017): 10.

12. Robin Dunbar, *Human Evolution* (New York: Penguin, 2014), 163.

13. Gowlett, Gamble, and Dunbar, "Human Evolution," 695–96.

14. Michael Tomasello, *The Cultural Origins of Human Cognition* (Cambridge, MA: Harvard University Press, 1999), loc. 39, Kindle.

15. James R. Hurford, *The Origins of Language: A Slim Guide* (Oxford: Oxford University Press, 2014), 68; Cheney and Seyfarth, *Baboon Metaphysics,* loc. 2408, Kindle: "Evidence for teaching by nonhuman primates...can be summarized by one word: scant."

16. Tomasello, *The Cultural Origins of Human Cognition,* loc. 5, Kindle: "Faithful social transmission...can work as a ratchet to prevent slippage backward—so that the newly invented artifact or practice preserves its new and improved form at least somewhat faithfully until a further modification or improvement comes along." Tomasello calls this *collaborative learning.*

17. Steven Pinker, *The Sense of Style: The Thinking Person's Guide to Writing in the Twenty-First Century* (New York: Penguin, 2015), 110.

18. This idea is suggested by Roth, *The Long Evolution of Brains and Minds,* 264; on the unique human capacity to remember many words, see Hurford, *The Origins of Language,* 119.

19. See Terrence Deacon, *The Symbolic Species: The Co-Evolution of Language and the Brain* (New York: W. W. Norton, 1998), and Michael Tomasello, *Why We Cooperate* (Cambridge, MA: MIT Press, 2009). For recent surveys of the evolution of language, see W. Tecumseh Fitch, *The Evolution of Language* (Cambridge: Cambridge University Press, 2010), and Peter J. Richerson and Robert Boyd, "Why Possibly Language Evolved," *Biolinguistics* 4, nos. 2/3 (2010): 289–306. Alex Mesoudi, *Cultural Evolution: How Darwinian Theory Can Explain Human Culture and Synthesize the Social Sciences* (Chicago: University of Chicago Press, 2011), is a fine recent survey of the rich body of research on cultural change from a Darwinian perspective.

20. Eric R. Kandel, *In Search of Memory: The Emergence of a New Science of Mind* (New York: W. W. Norton, 2006), loc. 330, Kindle.

21. William H. McNeill, "*The Rise of the West* After Twenty-Five Years," *Journal of World History* 1, no. 1 (1990): 2.

22. Sally McBrearty and Alison S. Brooks, "The Revolution That Wasn't: A New Interpretation of the Origin of Modern Human Behavior," *Journal of Human Evolution* 39 (2000): 453–563.

23. The image is from Peter J. Richerson and Robert Boyd, *Not by Genes Alone: How Culture Transformed Human Evolution* (Chicago: University of Chicago Press, 2005), 139.

24. Dunbar, *Human Evolution,* 13.

25. There is a good brief overview in Chris Scarre, ed., *The Human Past: World Prehistory and the Development of Human Societies* (London: Thames and Hudson, 2005), 143–45.
26. Peter Hiscock, "Colonization and Occupation of Australasia," in *Cambridge World History*, vol. 1, ed. Merry Wiesner-Hanks (Cambridge: Cambridge University Press, 2015), 452.
27. These migrations are described well in Peter Bellwood, *First Migrants: Ancient Migration in Global Perspective* (Malden, MA: Wiley-Blackwell, 2013).
28. On the early-dispersal model, see Hiscock, "Colonization and Occupation of Australasia," 433–38.
29. Figures from Christian, *Maps of Time*, 143.
30. Marshall Sahlins, "The Original Affluent Society," *Stone Age Economics* (London: Tavistock, 1972), 1–39.

Chapter 8. Farming: Threshold 7

1. Vaclav Smil, *Harvesting the Biosphere: What We Have Taken from Nature* (Cambridge, MA: MIT Press, 2013).
2. Jared Diamond, *Guns, Germs, and Steel: The Fates of Human Societies* (London: Vintage, 1998), develops the idea of a natural experiment in its final chapter.
3. See http://www.theaustralian.com.au/national-affairs/indigenous/aborigines-were-building-stone-houses-9000-years-ago/news-story/30ef4873a7c8aaa2b80d01a12680df77.
4. A fine recent overview of changing gender roles in human history is Merry E. Wiesner-Hanks, *Gender in History: Global Perspectives*, 2nd ed. (Malden, MA: Wiley-Blackwell, 2011).
5. Marc Cohen, *The Food Crisis in Prehistory* (New Haven, CT: Yale University Press, 1977), 65: "Groups throughout the world would be forced to adopt agriculture within a few thousand years of one another."
6. Chris Scarre, ed., *The Human Past: World Prehistory and the Development of Human Societies* (London: Thames and Hudson, 2005), 214–15.
7. Bruce Pascoe, *Dark Emu: Black Seeds: Agriculture or Accident?* (Broome, Australia: Magabala Books, 2014), describes many indigenous Australian cultivation techniques; the sickles are described at loc. 456, Kindle.

8. This is a central argument of Jared Diamond's wonderful *Guns, Germs, and Steel.*

9. Peter Bellwood, *First Migrants: Ancient Migration in Global Perspective* (Malden, MA: Wiley-Blackwell, 2013), 124.

10. Smil, *Harvesting the Biosphere,* loc. 2075, Kindle.

11. Merry Wiesner-Hanks, ed., *Cambridge World History*, vol. 2 (Cambridge: Cambridge University Press, 2015), 221, 224–28.

12. Robin Dunbar, *Human Evolution* (New York: Penguin, 2014), 77.

Chapter 9. Agrarian Civilizations

1. Richard Lee, "What Hunters Do for a Living, or, How to Make Out on Scarce Resources," in *Man the Hunter,* ed. R. Lee and I. DeVore (Chicago: Aldine, 1968).

2. Chris Scarre, ed., *The Human Past: World Prehistory and the Development of Human Societies* (London: Thames and Hudson, 2005), 403.

3. Cited in Alfred J. Andrea and James H. Overfield, *The Human Record: Sources of Global History,* vol. 1, 4th ed. (Boston: Wadsworth, 2008), 23–24.

4. Cited in Robert C. Tucker, ed., *The Marx-Engels Reader,* 2nd ed. (New York: W. W. Norton, 1978), 608.

5. Hans J. Nissen, "Urbanization and the Techniques of Communication: The Mesopotamian City of Uruk During the Fourth Millennium BCE," in *Cambridge World History,* vol. 3, Merry Wiesner-Hanks, ed. (Cambridge: Cambridge University Press, 2015), 115–16.

6. Mark McClish and Patrick Olivelle, eds., *The Arthasastra: Selections from the Classic Indian Work on Statecraft* (Indianapolis: Hackett Publishing, 2012), sections 1.4.13–15, Kindle.

7. Ibid., sections 1.4.1–1.4.4, 1.5.1.

8. Ibid., section 2.36.3.

9. Ibid., section 2.35.4.

10. Thomas Piketty, *Capital in the Twenty-First Century,* trans. Arthur Goldhammer (Cambridge, MA: Harvard University Press, 2014), 270, and see page 258, table 7.2.

Chapter 10. On the Verge of Today's World

1. Grace Karskens, *The Colony: A History of Early Sydney* (New South Wales: Allen and Unwin, 2009), loc. 756–79, Kindle.

2. The intensifying global hunt for new resources is described superbly in John Richards, *The Unending Frontier: Environmental History of the Early Modern World* (Berkeley: University of California Press, 2006).

3. Alfred W. Crosby, *Ecological Imperialism: The Biological Expansion of Europe, 900–1900* (Cambridge: Cambridge University Press, 1986).

4. Felipe Fernández-Armesto, *Pathfinders: A Global History of Exploration* (New York: W. W. Norton, 2007), 161 and following.

5. David Wootton, *The Invention of Science: A New History of the Scientific Revolution* (New York: Penguin, 2015), 68.

6. Cited in Steven J. Harris, "Long-Distance Corporations, Big Sciences, and the Geography of Knowledge," *Configurations* 6 (1998): 269.

7. Wootton, *The Invention of Science*, 37.

8. Ibid., 54.

9. Ibid., 35.

10. Ibid., 5–6, 8–9.

11. Margaret Jacob and Larry Stewart, *Practical Matter; Newton's Science in the Service of Industry and Empire, 1687–1851* (Cambridge, MA: Harvard University Press, 2004), 16.

12. David Christian, *"Living Water": Vodka and Russian Society on the Eve of Emancipation* (Oxford: Oxford University Press, 1990).

13. E. A. Wrigley, *Energy and the English Industrial Revolution* (Cambridge: Cambridge University Press, 2011), loc. 298–306, Kindle. Malthus, Jevons, Ricardo, and Mill also accepted that the natural world set limits to growth; see the discussion in Donald Worster, *Shrinking the Earth: The Rise and Decline of American Abundance* (Oxford: Oxford University Press, 2016), 44–49.

14. Alfred W. Crosby, *Children of the Sun: A History of Humanity's Unappeasable Appetite for Energy* (New York: W. W. Norton, 2006), 60.

15. Wrigley, *Energy and the English Industrial Revolution,* loc. 2112, Kindle.

16. On the history of the Newcomen engine and its links to the scientific revolution, see Wootton, *The Invention of Science,* chapter 14.

17. Wrigley, *Energy and the English Industrial Revolution,* loc. 2112, Kindle.
18. Daniel Yergin, *The Prize: The Epic Quest for Oil, Money, and Power* (New York: Free Press, 1991), chapter 1.
19. Ibid., 16.

Chapter 11. The Anthropocene: Threshold 8

1. Graham Allison and Philip Zelikow, *Essence of Decision: Explaining the Cuban Missile Crisis,* 2nd ed. (New York: Longman, 1999), 271.
2. Angus Maddison, *The World Economy: A Millennial Perspective* (Paris: Organisation for Economic Co-Operation and Development, 2001), 127.
3. Tim Lenton, *Earth Systems Science: A Very Short Introduction* (Oxford: Oxford University Press, 2016), 82.
4. Ha-Joon Chang, *Economics: The User's Guide* (New York: Pelican, 2014), 429, based on figures from the World Bank.
5. Lenton, *Earth Systems Science,* 82, 96–97.
6. The scientist was Wally Broecker. Cited in David Christian, "Anthropocene Epoch," in *The Berkshire Encyclopedia of Sustainability, Vol. 10: The Future of Sustainability,* ed. Ray Anderson et al. (Barrington, MA: Berkshire Publishing, 2012), 22.
7. Jan Zalasiewicz and Colin Waters, "The Anthropocene," in *The Oxford Research Encyclopedia, Environmental Science* (Oxford: Oxford University Press, 2015), 4–5.

Chapter 12. Where Is It All Going?

1. Kim Stanley Robinson's Mars trilogy—*Red Mars* (1993), *Green Mars* (1994), *Blue Mars* (1996)—offers a rich and vivid science-fictional account of what the colonization of Mars might look like.
2. Joseph Campbell, *The Hero with a Thousand Faces,* 2nd ed. (Princeton, NJ: Princeton University Press, 1968), 46.
3. J. S. Mill, "Of the 'Stationary State,'" in *The Principles of Political Economy,* Google Books, http://www.efm.bris.ac.uk/het/mill/book4/bk4ch06.
4. Johan Rockström et al., "A Safe Operating Space for Humanity," *Nature* 461 (September 24, 2009): 472–75; updated in Will Steffen et al., "Planetary Boundaries: Guiding Human Development on a Changing Planet," *Science* (January 2015): 1–15.

5. Steffen et al., "Planetary Boundaries," 1.
6. The idea of a mature Anthropocene is explored in David Grinspoon, *Earth in Human Hands: Shaping Our Planet's Future* (New York: Grand Central Publishing, 2016). I have borrowed some of the ideas in this section from Paul Raskin, *Journey to Earthland: The Great Transition to Planetary Civilization* (Boston: Tellus Institute, 2016).
7. I've taken details of the following account from Sean Carroll's wonderful book *The Big Picture: On the Origins of Life, Meaning, and the Universe Itself* (New York: Dutton, 2016), loc. 878, Kindle.

Index

A

absolute zero, 23, 37

absorption lines, 33, 55

Abu Hureyra, 198–99

accretion, 63–64, 65, 66, 100, 104

Acheulean axes, 168

activation energy, 47, 61, 88, 93, 294

Adams, Douglas, 91

adaptation, 81, 83

adaptive radiations, 132–33, 134–36, 142, 151

affluent foragers, 196–98, 199, 205–6

Afro-Eurasia, 194–95, 201, 208, 229–32, 238–40, 242

agrarian civilizations, 210–35, 237, 251, 278; cities in, 214, 217, 218–21; energy flows in, 213, 219–20, 221, 224, 225–26, 232–35; inequality in, 214–15, 220; leaders and power structures in, 215–16, 217–18, 220–32; measuring change in, 232–35; mobilization in, 221–25; new technologies in, 213, 222, 229–30; religions in, 214, 215, 219, 230, 232; social structures in, 211, 212, 215;

specialization in, 212–14, 217; surplus wealth in, 212; trade in, 213–14, 229, 231; transportation in, 229, 231; underclasses in, 210–11, 212

algae, 118, 129, 130

Alpher, Ralph, 36–37

Alvarez, Walter, 10, 148–49

Americas, 102, 193, 195, 201, 203, 208, 232, 237; European navigators' crossings to, 240–41, 242–43; global-exchange networks and, 244–45; humans' arrival in, 181, 183, 184, 186, 194

amino acids, 57, 61, 87, 88, 93, 95–96

amniotes, 141–43. *See also* mammals

amphibians, 138

Andromeda, 34, 303–4

angiosperms, 150

animals, 79, 103, 122, 131, 138; domestication of, 190–91, 192, 196–97, 198, 200, 201, 207–8, 276; earliest multicellular, 125, 132, 133–34; human impact on, 276–77; spread onto land, 127, 135, 137, 138

salt, 59–60
Sarich, Vincent, 161
Saturn, 63, 66–67, 86
Schrödinger, Erwin, 26
Schumpeter, Joseph, 243
science, 246, 247; in future,
    302–3; Paleolithic, 185–87.
    See also technologies, new
Search for Extraterrestrial
    Intelligence (SETI), 80
sea slugs, 159
Sedgwick, Adam, 133
seismographs, 67
sensors, 79, 113–14, 144–45, 146
sexual reproduction, 123–24
Shapley, Harlow, 39
Sherratt, Andrew, 207–8
Siberian Traps, 136
silicon, 65
slavery and forced labor, 210, 212,
    215, 216, 244, 250, 261, 272, 282
Slipher, Vesto, 33, 34
Smith, Adam, 236, 248, 251, 295
Smith, William, 68
sociability, 160, 165, 167, 174;
    villages of agrarian era and,
    204–7, 209
sodium, 59–60
solar system, 61, 70; formation of,
    62–64; location of, 64–65, 86
solar wind, 62–63, 65
sonar technology, 103
space-time, 19, 265, 304, 305
specialization, 212–14, 215, 217
spectroscopes, 33, 55, 57, 60
stability, big life and, 126
Standard of Ur, 220
stars, 56, 76–77, 86, 95, 304–5;
    Cepheid variables, 34–36, 52;

estimating distance to, 33–34;
    formation of, 39, 42–45, 47, 56,
    304; held together by energy
    flows, 40; life histories of, 46,
    49–55; luminosity (brightness)
    of, 34, 49, 50, 55, 61; movement
    of, 32–33, 35–36, 61; plasmas
    in centers of, 30; supernovas,
    52–53; surface temperature
    of, 55
states, agrarian, 221–32;
    emergence of, 216, 217–18;
    mobilization in, 218, 221–25;
    modern states vs., 217–18;
    spread of, 227–32; writing and,
    219–20, 226, 229
steady-state theory, 18, 35, 36
steam engines, 253–54, 255, 256,
    264
Steno, Nicolas, 68
stereoscopic vision, 158–59
Stockholm Resilience Centre,
    297
stomata, 137
stromatolites, 89, 116, 126, 127
strong nuclear force, 23, 24,
    29, 58
Suess, Eduard, 99, 101
sun, 47, 50, 51, 55, 57, 65, 66, 126,
    152; formation of, 62; future
    of, 303; heat at Earth's surface
    from, 106, 107, 109; parallax
    method and, 33–34; solar wind
    of, 62–63; ultraviolet radiation
    from, 118
sunlight, photosynthesis and,
    115–19, 122, 131
supercontinents, 101–2, 104–5,
    130, 240, 303